2nd International Conference on Chemical Materials and Process (ICCMP 2016)

MATEC Web of Conferences
Volume 64 (2016)

Copenhagen, Denmark
11-13 May 2016

Editors:

O. Adiguzel
T. Zhi-Gang
S. Wongkasemjit
C. C. Manole

ISBN: 978-1-5108-2601-4

Printed from e-media with permission by:

Curran Associates, Inc.
57 Morehouse Lane
Red Hook, NY 12571

Some format issues inherent in the e-media version may also appear in this print version.

Printed by Curran Associates, Inc. (2016)

For additional information, please contact EDP Sciences – Web of Conferences at the address below.

EDP Sciences – Web of Conferences
17, Avenue du Hoggar
Parc d'Activité de Courtabœuf
BP 112
F-91944 Les Ulis Cedex A
France

Phone: +33 (0) 1 69 18 75 75
Fax: +33 (0) 1 69 28 84 91

contact-edps@webofconferences.org

Additional copies of this publication are available from:

Curran Associates, Inc.
57 Morehouse Lane
Red Hook, NY 12571 USA
Phone: 845-758-0400
Fax: 845-758-2633
Email: curran@proceedings.com
Web: www.proceedings.com

TABLE OF CONTENTS

Message from the General Chair

Dear Delegates,

Welcome to the ICCCP 2016, the event of Sponsored by China and Asia-Pacific Chemical, Biological & Environmental Engineering Society.

The year 2016, being the 7th anniversary of the formation of APCBEES, bears a special significance for us. The society which took off during Feb 2009, has remained airborne & completed 7 years of its journey, vide the support of all of its stakeholders. The journey was indeed, challenging but extremely educative and satisfying too. Challenging because the task of establishing a completely new society & using it as an umbrella for bringing very focused personnel under it, was certainly not an easy one. And to be frank, the journey was not entirely free of turbulence either. At the same time, the responsibility has taught us, among others, to cultivate patience and not to expect any radical solution of any problem overnight. Ultimately the feedback, informing us about our being able to meet the expectations of the stakeholders, is most satisfying and motivating. I solicit your continued patronage in the coming days & your views, stating both the merits & demerits of our style of working, will keep us navigating in the right direction.

Coming to ICCCP 2016, after carrying out an array of activities in Asia over the last 7 years, we decided to make the presence of APCBEES felt in other corners of the world too. The aim was to make the society a national outfit with global outlook. Accordingly, after celebrating the 7th anniversary in Hong Kong, we selected the picturesque Roma as the place for organizing our ICCCP this year.

The entire team of ICCCP as well as the Members of the ExeCom of APCBEES had promoted the event vigorously. I wish to express my sincere gratitude to all of them. The Program Chairs & the IPC members demonstrated their extreme professionalism by completing the process of reviewing and short listing of papers within the stipulated time.

We would like to thank our outstanding keynote speakers Prof. Jacek Namieśnik, Prof. Hartmut Hinz, Prof. Marco Casini, Prof. Michael Schmidt, for sharing their deep insights on future challenges and trends in Chemistry and Chemical Process. Also with our special thanks to Associate Prof. Chi-wai KAN and Prof. Tuomas Koiranen who are unable to attend the conference, but give us great support. Thanks for Technical Committee member Assoc. Prof. Ozgur Ceylan and Dr. Mohamed Elhefnawy, who coming to our conference and will share their research topics in our conference.

I wish to express my heartfelt thanks to all the authors who submitted their research findings for presentation at ICCCP 2016. Various constraints, including putting more emphasis on quality, compelled us to accept only a subset of the total no. of submitted manuscripts. My hearty congratulations go to all the authors of the accepted papers. I am sure this acceptance will make them pursue their research with more vigor.

Apart from usual participation which involves renewing old friendships & building new bridges, I believe that all of you will find your sojourn in Bali comfortable and the its scenic beauty enjoyable. I also hope that we will continue to be in touch in the coming days. Your feedback – good or bad – will be one of the essential ingredients for our future success & I urge you once more to assist us towards that. Before ending, let me wish all of you a very happy & prosperous 2016.

General Chair, ICCCP 2016

Committee Members

Conference General Co-Chairs

Prof. Jacek Namieśnik
Faculty of Chemistry, Gdansk University of Technology, Poland

Program Co-Chairs

Associate Prof. Chi-wai KAN
The Hong Kong Polytechnic University, Hong Kong

Contact Chair

Ms Flora Feng
Asia-Pacific Chemical, Biological & Environmental Engineering Society

Technical Committee

Prof. Tuomas Koiranen
Lappeenvanta University of Technology, Finland

Prof. JERZY SĘK
Lodz University of Technology, Poland

Prof. Junkuo Gao
Zhejiang Sci_Tech University, China

Prof. Changjun Hou
Sichuan University of Science &Engineering, Zigong, China

Prof. Elena Ivashkina
Tomsk Polytechnic University, Tomsk, Russia

Assoc. Prof. Javeed A. Awan
Technical University of Denmark, Denmark

Assoc. Prof. Ozgur Ceylan
Muğla Sıtkı Koçman University, Turkey

Assoc. Prof. Khairul Sozana Nor Kamarudin
Universiti Teknologi Malaysia, Malaysia

Dr. Rodrigo J. Alvarez-Mendez
Universidad de Xalapa, Mexico

Dr. Mohamed Elsayed Elhefnawy
King Abdulaziz University, Saudi Arabia

Dr. Manel Wakkel
Institut National des Sciences Appliquées, Toulouse, France

Statement of Peer review

In submitting conference proceedings to MATEC Web of Conferences, I certify to the Publisher that I adhere to the **Policy on Publishing Integrity** of the journal in order to safeguard good scientific practice in publishing.

1. All articles have been subjected to peer review administered by the proceedings editors.
2. Reviews have been conducted by expert referees, who have been requested to provide unbiased and constructive comments aimed, whenever possible, at improving the work.
3. Proceedings editors have taken all reasonable steps to ensure the quality of the materials they publish and their decision to accept or reject a paper for publication has been based only on the merits of the work and the relevance to the journal.

Title, date and place of the conference

2016 6th International Conference on Chemistry and Chemical Process
February 1-3, 2016 in Rome, Italy

Proceedings editor(s):

Prof. Jacek Namiesnik, chemanal@pg.gda.pl
Faculty of Chemistry, Gdansk University of Technology, Poland
Prof. Chi-wai KAN, tccwk@polyu.edu.hk
The Hong Kong Polytechnic University, Hong Kong

Date and editor's signature Dr. Dan Yang, CBEES, Hong Kong

MATEC Web of Conferences **49**, 01001 (2016)
DOI: 10.1051/matecconf/20164901001

Modeling of Benzene with Ethylene Alkylation

Elena Khlebnikova [1, a], Irena Dolganova [1], Elena Ivashkina [1], Stanislav Koshkin [1, 2]

[1] *Tomsk Polytechnic University, Tomsk, Russia*
[2] *LLC NIOST, Tomsk, Russia*

Abstract. The paper considers the benzene alkylation with ethylene model development with the use of zeolite catalyst. A list of reactions occurring in the alkylation reactor was made and the thermodynamic possibility of determination of these reactions by the change of Gibbs energy was defined. The paper presents the hydrocarbons transformation scheme, which includes the grouping of components on the basis of their reactivity and the degrees of compensation values of the corresponding reactions. Drawing on the obtained data the authors developed the kinetic model of the alkylation of benzene with ethylene.

1 Introduction

Since the 70s of the last century it became possible to produce ethylbenzene with the use of heterogeneous zeolite catalyst on a commercial scale. Heterogeneous catalytic process gradually started to replace the old homogeneous catalytic technology. Nowadays most of the existing ethylbenzene production plants operate using the zeolite catalyst [1, 2]. According to the data of 2014 total world ethylbenzene production capacity is around 37 millions tons per year [3].

Technology of ethylbenzene production includes the stages of alkylation and transalkylation. The control of technological parameters such as benzene-to-ethylene ratio in the feed and temperature lets to conduct the process under the optimum conditions and monitor the output of ethylbenzene [4, 5].

Although the technology developers provide their recommendations on the values of these parameters, the development of the model of ethylbenzene synthesis process and its subsequent modeling aimed at the increase of the resource and energy efficiency of the production plant is one of the most feasible ways of customization of the process to particular cases [6].

2 Object of the research

The ethylbenzene technology is developed for ethylbenzene production with the use of the liquid phase and zeolite catalyst.

The industrial plant operates continuously and uses 2 different reactors for alkylation and transalkyation.

In alkylation reactor ethylene in its liquid phase goes through almost a complete reaction accompanied by the excess of benzene, which is necessary to form ethylbenzene (EB):

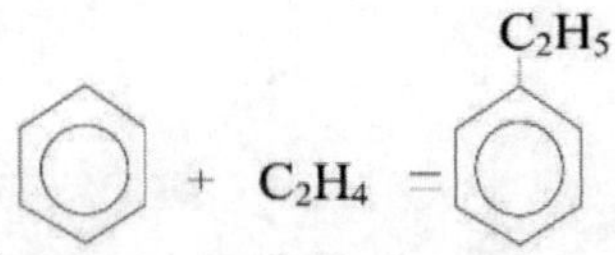

benzene ethylene EB

The process takes place at T = 220 - 255 °C and P=3.4 MPa. This reaction is exothermic, which leads to gradual temperature increase from one catalyst bed to the following one. This is the reason why alkylation reactors are typically equipped with intermediate coolers for keeping the total temperature rise.

Then ethylbenzene converts to produce small amounts of polyethylbenzenes (PEB):

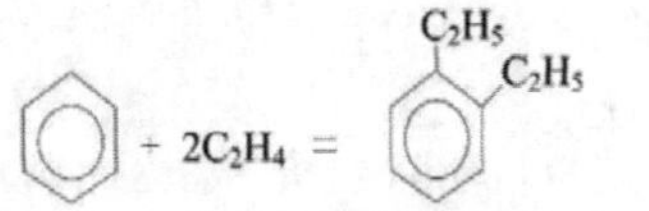

benzene ethylene diethylbenzene

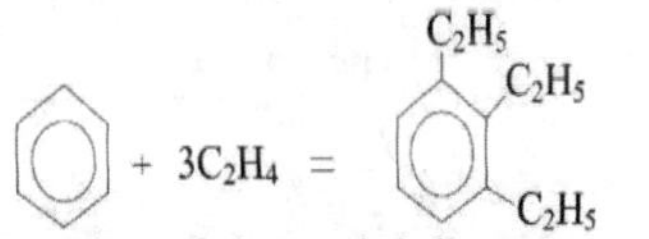

benzene ethylene triethylbenzene

These reactions are the main side ones and generate considerable quantities of heat being exothermic ones.

PEB are separated from ethylbenzene and then converted to EB in the transalkyation reactor at T = 190 - 220 °C and P = 3 - 3.5 MPa.

By now the chemistry of ethylbenzene synthesis is quite well understood [7, 8]. In addition to the main reactions leading to ethylbenzene formation it's known a number of side reactions. Altogether these reactions are a challenging task from the perspective of mathematical description and its subsequent use. Representation and adequacy of a mathematical model and its ability to be

[a] Corresponding author: elena.khle@gmail.com

approximate enough to the real data depend on reaction scheme of the model. This scheme should be quite simple, but at the same time accurate enough to describe the process [9].

3 Model development

Reactions that occur in alkylation and transalkyation processes and that are based on different types of the zeolite catalyst were investigated by certain authors [10-12]. However, alkylation and transalkyation have different conversion processes schemes, and at the same time they have quite much in common. So the main task is to determine the pathway of the development of secondary reactions: transalkyation, disproportionation, dealkylation, cracking and other ones, the result of which is the formation of by-products. Some steps were taken to make up a detailed reaction scheme based on industrial data and the existing knowledge about the process [13]. We offer the method of designing of the math model that combines industrial data, research results and quantum-chemical calculation.

The list of possible reactions for benzene alkylation was made with account of the quality of sources (ethylene and benzene), product composition and aspects of the process chemistry.

Thermodynamic possibility of these reactions has been confirmed by determination of the isobaric-isothermal parameter – the change of Gibbs energy ΔG at T=510 °K and P=3.4 MPa. As for the target reaction Gibbs energy is equal to -43.4 kJ / mol and according to all defined ΔG of the considered reactions, thermodynamic possibility of these reactions is probable. The determination was performed with the use of *Gaussian-98* program by PM3 semi-empirical method.

The graphical description of the reactions pathway of benzene alkylation is shown in Figure 1. Among them there are benzene alkylation of ethylene and propylene (ethylene feed impurity), toluene (benzene feed impurity) alkylation, next step of alkylation caused by the PEB formation, heavy by-product formation (diphenylethane) through the intermediate formation in the process of incomplete benzene alkylation.

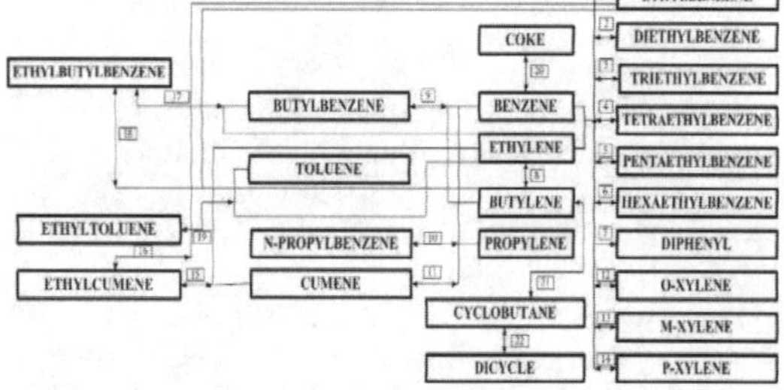

Figure 1. Detailed scheme of hydrocarbons conversion in the process of alkylation of benzene with ethylene

The scheme needs to be formalized (simplified) in order it could be used for future computations; at the same time it is to provide the highly accurate description of the process.

Formalization was performed by means of uniting individual substances into the pseudo-components having similar reactivity in respect of a particular reaction. The determination of compensation degree, which is characterized by the partial compensation of the energy expended on the breaking of the old bonds by the energy of the new bonds formation on the reaction route, was considerably helpful in solving the problems of catalytic action prediction.

Therefore, further step is to study the degree of compensation for the list of considered reactions.

$$H = \frac{\sum D_i - E_a}{\sum D_i} \cdot 100\% \qquad (1)$$

where $\sum D_i$ – the energy of bounds breaking, kJ / mol, E_a – energy of activation, kJ / mol.

The main problem in calculation of the compensation energy is to determine the activation energy of the reaction.

As the first approximation, the activation energy was calculated according to Polanyi-Semenov equation for exothermic reactions [14].

$$E_a = 11.5 + 0.25 \cdot \Delta H \qquad (2)$$

where E_a – energy of activation, kcal / mol, ΔH – heat of reaction, kcal / mol.

Enthalpy of the reactions was calculated with the use of quantum-chemical methods of thermodynamic functions determination. The calculated enthalpy of alkylation reaction of benzene with ethylene (-117.9 kJ / mol) was compared with the literature data (-113 kJ / mol) [15] as a proof of the relevance of the method for the unknown enthalpy of some reactions in the literature. All the considered reactions are exothermic.

Then the activation energy of the given reaction was calculated in first approximation with the use of the empiric equation. The value of the reaction mentioned above is 78.5 kJ / mol in comparison with 63.4 kJ / mol in [16]. It's important to note that activation energy that was found for all reaction refers to the non-catalytic process, and it was good for the first approximation on next stage of the parameter estimation. The meanings of defined activation energy according to Polanyi-Semenov equation at T=510 °K and P=3.4 MPa are represented in Table 1. However, it might be improved with the use of literature sources of activation energy values for some reactions. After that the activation energy could be calibrated to the activation energy of the reaction.

The next step in calculating the degree of compensation is the determination of the energy of bonds breaking. Let's consider the methods based on the example of the above-mentioned reaction.

To form ethylbenzene it's necessary to break the two bonds: C-H bond in a molecule of benzene and C-H bond in a molecule of ethylene. The energies of these bonds have been calculated with the use of *Gaussian-98*.

The energy of the first bond was 433.8 kJ / mol, the energy of the second one was 215.5 kJ / mol. The energy of bonds breaking was equal to 649.3 kJ / mol. Similarly, the energy of bonds breaking was calculated for the rest of the reactions, and then the degree of compensation was calculated by the formula (1). The results are shown in Table 1. The literature values of the energy of bonds breaking and their comparison with the determined values are also shown in Table 1.

Table 1. Results of determined degrees of reaction compensation

Reaction	E_a (Polanyi-Semenov), kJ / mol	$\sum D_i$, kJ / mol	$\sum D_i$, kJ / mol (literature)	$\Delta\sum D_i$, %	H, %
Benzene + Ethylene = Ethylbenzene	78.5	649.3	715.0	9.2	90.2
Ethylbenzene + Ethylene = Diethylbenzene	86.1	633.5	715.0	11.4	92.6
Diethylbenzene + Ethylene = Triethylbenzene	90.7	617.7	715.0	13.6	88.9
Triethylbenzene + Ethylene = Tetraethylbenzene	84.4	602.0	715.0	15.8	89.1
Tetraethylbenzene + Ethylene = Pentaethylbenzene	84.8	586.2	715.0	18.0	88.8
Pentaethylbenzene + Ethylene = Hexaethylbenzene	93.2	570.4	715.0	20.2	87.9
2Benzene + Ethylene= Diphenylethane + Hydrogen	94.1	1083.2	1100.0	1.5	93.6
Ethylene + Ethylene = Butylene	77.7	882.2	955.0	7.6	92.8
Benzene + Butylene = Butylbenzene	105.8	762.9	685.0	11.4	90.3
Benzene + Propylene = n-Propylbenzene	99.1	788.2	685.0	15.1	90.9
Benzene + Propylene = Cumene	107.9	788.2	685.0	15.1	90.5
Benzene + Ethylene = o-Xylene	63.8	1298.7	1428.0	9.1	95.6
Benzene + Ethylene = m-Xylene	55.0	1298.7	1428.0	9.1	95.8
Benzene + Ethylene = p-Xylene	55.0	1298.7	1428.0	9.1	95.8
Cumene + Ethylene = Ethylcumene	94.5	633.0	715.0	11.5	89.0
Ethylbenzene + Propylene = Ethylcumene	123.5	772.3	715.0	8.0	89.5
Butylbenzene + Ethylene = Ethylbutylbenzene	92.8	635.4	715.0	11.1	89.1
Ethylbenzene + Butylene = Ethylbutylbenzene	120.5	747.1	715.0	4.5	89.3
Toluene + Ethylene = Ethyltoluene	108.4	747.6	715.0	4.6	89.9
6Benzene = Coronene + Hydrogen	1179.8	5206.0	4980.0	4.5	90.3
Butylene = cis-Butene	183.1	952.5	1014.0	6.1	89.0
cis-Butene + cis-Butene = Dicyclobutane + Hydrogen	166.3	809.6	830.0	2.5	87.8

Computation results show the good comparability of the determined values of energy of bonds breaking with the literature data.

4 Results and discussions

In order to formalize the transformation schemes, reactions were divided into two groups by the criterion of the similarity of their values of compensation degree (Table 2). The degree of compensation for the same group of reactions differs approximately by 1% [17-19].

Table 2. Alkylation reactions grouping by their degrees of compensation

Reaction group	Average H, % for reaction group
1. Ethylbenzene formation	90.2
Benzene + Ethylene = Ethylbenzene	
2. Diethylbenzene formation	92.6
Ethylbenzene + Ethylene = Diethylbenzene	
3. Triethylbenzene formation	88.9
Diethylbenzene + Ethylene = Triethylenebenzene	
4. Heavy compounds formation	89.8
Triethylbenzene + Ethylene = Polyethylenebenzene	
2Benzene + Ethylene = Diphenylethane + Hydrogen	
5. Other monoalkylates formation	90.6
Benzene + Butylene = Monoalkylate	
Benzene + Propylene = Monoalkylate	
6. Xylene formation	95.7
Benzene + Ethylene = Xylene	
7. Other dialkylates formation	89.4
Ethylbenzene + Propylene = Dialkylate	
Ethylbenzene + Butylene = Dialkylate	
Monoalkylate + Ethylene = Dialkylate	
Toluene + Ethylene = Dialkylate	
8. Coke formation	90.3
6Benzene = Coke + Hydrogen	
9. Cyclo and dicyclo formation	88.4
Butylene = cis-Butene	
cis-Butene + cis-Butene = Dicyclobutane + Hydrogen	
10. Butylene formation	92.8
Ethylene + Ethylene = Butylene	

Thus, it is possible to identify the number of pseudo-components: the heavy compounds which are composed of individual components such as tetra-, penta-, hexaethylbenzene and diphenylethane; monoalkylates (butylbenzene, n-propylbenzene, cumene), dialkylates (ethylcumene, ethylbutylbenzene), xylenes (o-, m-, p-xylene), cyclic and dicyclic hydrocarbons. Formalized scheme of hydrocarbons transformations is shown in Figure 2.

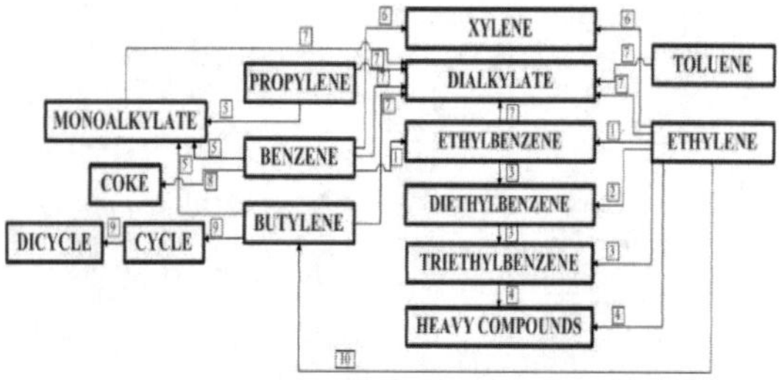

Figure 2. Formalized scheme of hydrocarbons transformations in benzene in process of its alkylation with ethylene

According to the law of mass action the expressions of the rates of alkylation and transalkylation reactions can be put as follows (Table 3).

Table 3. Reaction rate expressions in alkylation and transalkylation

Reaction group	Reaction rate expression
1. Ethylbenzene formation	$W_1 = k_{0(1)} \cdot e^{-Ea(1)/RT} \cdot C_{benzene} \cdot C_{ethylene}$
2. Diethylbenzene formation	$W_2 = k_{0(2)} \cdot e^{-Ea(2)/RT} \cdot C_{ethylbenzene} \cdot C_{ethylene}$
3. Triethylbenzene formation	$W_3 = k_{0(3)} \cdot e^{-Ea(3)/RT} \cdot C_{diethylbenzene} \cdot C_{ethylene}$
4. Heavy compounds formation	$W_{4.1} = k_{0(4)} \cdot e^{-Ea(4.1)/RT} \cdot C_{triethylbenzene} \cdot C_{ethylene}$ $W_{4.2} = k_{0(4)} \cdot e^{-Ea(4.2)/RT} \cdot C^2_{benzene} \cdot C_{ethylene}$
5. Other monoalkylates formation	$W_{5.1} = k_{0(5)} \cdot e^{-Ea(5.1)/RT} \cdot C_{benzene} \cdot C_{butylene}$ $W_{5.2} = k_{0(5)} \cdot e^{-Ea(5.2)/RT} \cdot C_{benzene} \cdot C_{propylene}$
6. Xylene formation	$W_6 = k_{0(6)} \cdot e^{-Ea(6)/RT} \cdot C_{benzene} \cdot C_{ethylene}$
7. Other dialkylates formation	$W_{7.1} = k_{0(7)} \cdot e^{-Ea(7.1)/RT} \cdot C_{ethylbenzene} \cdot C_{propylene}$ $W_{7.2} = k_{0(7)} \cdot e^{-Ea(7.2)/RT} \cdot C_{ethylbenzene} \cdot C_{butylene}$ $W_{7.3} = k_{0(7)} \cdot e^{-Ea(7.3)/RT} \cdot C_{monoalkylate} \cdot C_{ethylene}$ $W_{7.4} = k_{0(7)} \cdot e^{-Ea(7.4)/RT} \cdot C_{toluene} \cdot C_{ethylene}$
8. Coke formation	$W_8 = k_{0(8)} \cdot e^{-Ea(8)/RT} \cdot C^6_{benzene}$
9. Cyclo and dicyclo formation	$W_{9.1} = k_{0(9)} \cdot e^{-Ea(9.1)/RT} \cdot C_{butylene}$ $W_{9.2} = k_{0(9)} \cdot e^{-Ea(9.2)/RT} \cdot C^2_{cis\text{-}butene}$
10. Butylene formation	$W_{10} = k_{0(10)} \cdot e^{-Ea(_{10})/RT} \cdot C^2_{ethylene}$

where W_i – rate of i – group reaction; $k_{0(i)}$ – pre-exponential factor for i – group reaction; $Ea_{(i,j)}$ – activation energy of j - subgroup for i – group reaction.

Similar work was carried out with the transalkylation process. Kinetic models of alkylation and transalkylation are presented in Table 4.

Table 4. Kinetic models of alkylation and transalkylation

The kinetic model of alkylation	**The kinetic model of transalkylation**
$\begin{cases} \frac{dC_{benzene}}{dt} = -W_1 - 2W_{4.2} - W_{5.1} - W_{5.2} - \\ -W_6 - 6W_8 \\ \frac{dC_{ethylene}}{dt} = -W_1 - W_2 - W_3 - W_{4.1} - \\ -W_{4.2} - W_6 - W_{7..3} - W_{7.4} - 2W_{10} \\ \frac{dC_{butylene}}{dt} = -W_{5.1} - W_{7.2} - W_{9.1} + W_{10} \\ \frac{dC_{propylene}}{dt} = -W_{5.2} - W_{7.1} \\ \frac{dC_{cis\text{-}butene}}{dt} = W_{9.1} - 2W_{9.2} \\ \frac{dC_{dicyclobutane}}{dt} = W_{9.2} \\ \frac{dC_{hydrogen}}{dt} = W_8 + W_{4.2} + W_{9.2} \\ \frac{dC_{ethylbenzene}}{dt} = W_1 - W_2 - W_{7.1} - W_{7.2} \\ \frac{dC_{diethylbenzene}}{dt} = W_2 - W_3 \\ \frac{dC_{triethylbenzene}}{dt} = W_3 - W_4 \\ \frac{dC_{heavy\_compounds}}{dt} = W_{4.1} + W_{4.2} \\ \frac{dC_{monoalkylate}}{dt} = W_{5.1} + W_{5.2} - W_{7.3} \\ \frac{dC_{dialkylate}}{dt} = W_{7.1} + W_{7.2} + W_{7.3} + W_{7.4} \\ \frac{dC_{xylene}}{dt} = W_6 \\ \frac{dC_{toluene}}{dt} = -W_{7.4} \\ \frac{dC_{coke}}{dt} = W_8 \end{cases}$	$\begin{cases} \frac{dC_{benzene}}{dt} = -W_1 - W_2 - W_3 - W_{4.1} - \\ -W_{4.2} - W_{5.1} - 6W_7 - 2W_8 \\ \frac{dC_{ethylene}}{dt} = W_8 \\ \frac{dC_{diethylbenzene}}{dt} = -W_1 + W_2 \\ \frac{dC_{triethylbenzene}}{dt} = -W_2 + W_3 \\ \frac{dC_{polyethylbenzene}}{dt} = -W_3 \\ \frac{dC_{ethylbenzene}}{dt} = 2W_1 + W_2 + W_3 + 2W_{4.1} + W_{5.1} \\ \frac{dC_{butylbenzene}}{dt} = -W_{4.1} - W_{4.2} \\ \frac{dC_{monoalkylate}}{dt} = 2W_{4.2} + W_{5.1} \\ \frac{dC_{dioalkylate}}{dt} = -W_{5.1} - W_{5.2} \\ \frac{dC_{hydrogen}}{dt} = -W_{5.2} + W_6 + W_7 + W_8 \\ \frac{dC_{xylene}}{dt} = W_{5.2} \\ \frac{dC_{methane}}{dt} = W_{5.2} \\ \frac{dC_{cis\text{-}butene}}{dt} = -2W_6 \\ \frac{dC_{dicyclobutane}}{dt} = W_6 \\ \frac{dC_{diphenylethane}}{dt} = W_8 \\ \frac{dC_{coke}}{dt} = W_7 \end{cases}$

Initial conditions are: t=0, C_i=C_{0i}, where i – corresponding hydrocarbon.

The kinetic models presented in the paper are to become the basis for the computer simulation of the ethylbenzene production system development.

5 Conclusion

Thermodynamic analysis of alkylation and transalkylation reactions were performed with the use of quantum-chemical methods of calculating thermodynamic functions. The method of formalization of the hydrocarbons transformation scheme, which consists in the grouping of components by their reactivity drawing on the compensation degree values of the corresponding reactions, was developed. The correctness of the determination of the energy of bonds breaking in the reagent's molecules is confirmed by comparison of its results with the literature data. The kinetic models of alkylation of benzene with ethylene and polialkylbenzenes transalkylation were developed on the basis of the formalized transformations schemes.

References

1. Thomas F. Degnan Jr, C Morris Smith, Chaya R, Venkat, Appl. Catal. A: Gen. J. **221,** 283-294 (2001)
2. C.Perego, P. Ingallina, Catal. Today, **73**, 3-22 (2002)
3. Jennifer Carvajal-Diaz, David Byrne CEH (2015)
4. Robert A. Meyers, Handbook of Heterogeneous Catalysis, McGRAW-HILL, USA, 665 (2005)
5. N. Ebrahimi Ali, Z. Sharak Ashkan, A. Mousavi Seyyed, Faeze Aghazadeh, Ahmad Soltani, Chemical Engineering and Processing, **50**, 31-36 (2011)
6. Xiaoming Chen, Shiping Huang, Dapeng Cao Fluid.Ph.Equilibria, **260**, 146-152 (2007)
7. V.A. Fetisova, E.D. Ivashkina, E.N., Ivanchina, A.V. Kravtsov, Catal. In Industry 2-1, 55-61 (2010)
8. T.Odedairo, S. Al-Khattaf, Chemical Engineering J. **167**, 240-254 (2011)
9. Evgeniya Frantsina, Elena Ivashkina, Emiliya Ivanchina, Rostislav Romanovskii, Chemical Engineering Journal (available online 28 February 2015)
10. L. Forni, G. Cremona, F. Missineo, G. Bellusi, C. Perego, G. Pazzuconi, App. Cat. A:General J. **121**, 261-272 (1995)
11. F. Canavi, MCorazzari, E. Bencini, G.Goffredi, App. Cat. A:General J. **226**, 31-40 (2004)
12. Tseng-Chang Tsai, Shang-Bin Liu, Ikai Wang, App. Cat. A:General J. **181**, 355-398 (1999)
13. N. Hamendi, D. Iranshahi, M.R. Rahimpour, S. Raeissi, H. Rajaei, Journal of the Taiwan Inst. Of Chem. Eng J. **48**, 56-72 (2015)
14. Anthony G. Dixon, Advances in Chemical Engineering, USA, 301 (2014)
15. Kgutso Mokaema Alkyl-Transfer (Transalkylation) reactions of alkylaromatics on solid acid catalyst, Johannesburg, 68 (2005)
16. Hamid Ganji, Jafar S, Ahari et all, Petroleum&Coal J. **46**, 1, 55-63 (2004)
17. Howard F. Rase, Handbook of Commercial Catalyst. Heterogeneous Catalyst, CRC Press, USA, 551 (2000)
18. Nurmakanova A. E., Salishcheva A. A., Chudinova A. A. , Syskina A. A., Ivashkina Procedia Chemistry, **10**, 284-288 (2014)
19. Khlebnikova E. S., Bekker A. V., Ivashkina E. N., Procedia Chemistry, **10**, 297-304 (2014)

MATEC Web of Conferences **49**, 01002 (2016)
DOI: 10.1051/matecconf/20164901002

Kinetics and Mechanism of the Reaction of Coherently Synchronized Oxidation and Dehydrogenation of Cyclohexane by Hydrogen Peroxide

S. Aghamammadova [1], I. Nagieva [2], L. Gasanova [1], T. Nagiev [1, 2, a]

[1] *Nagiev Institute of Catalysis and Inorganic Chemistry, National Academy of Sciences of Azerbaijan*
[2] *Baku State University, Azerbaijani Republic*

Abstract. Inducing effect of hydrogen peroxide on synchronous oxidation reaction is accompanied by the occurrence of two interconnected and interacting reactions. The reaction of H_2O_2 decomposition (primary) generates the leading active $^{\bullet}OH$ and $HO_2^{\bullet}$ free radicals in a system. During the interaction of active and free radicals with the substrate, the conversion of the substrate occurs in the secondary reaction, coherently-synchronized with the primary one.
The mechanism of such coherently synchronized reactions is being examined in the process of cyclohexane oxidation with hydrogen peroxide in homogeneous and heterogeneous systems.
The process of coupling dehydrogenation of cyclohexane to cyclohexene and cyclohexadiene with hydrogen peroxide was carried out without catalyst. This process was also carried out at a fairly low temperature in a heterogeneous catalytic system.
Biomimetic catalyst, which simulates the basic functions of the enzyme group of oxidoreductase - catalase and monooxygenase, was used as the catalyst here. Characterised by its highly active and selective action these biomimetic catalysts are synthesized based on iron-porphyrin complexes simulating the active component of cytochrome P-450.
Based on this experimental researches, the complex reaction, consisting of parallel-sequential oxidation and dehydrogenation reactions, which are coherently synchronized, proceeds during the process of cyclohexane oxidation with biomimetic catalyst. Depending on the reaction parameters it is possible to deliberately adjust the direction of oxidation reaction and reaction rate.

1 Introduction

Inducing effect of hydrogen peroxide on synchronous oxidation reaction is accompanied by the occurrence of two interconnected and interacting reactions [1, 2]. The reaction of H_2O_2 decomposition (primary) generates the leading active $^{\bullet}OH$ and $HO_2^{\bullet}$ free radicals in a system. During the interaction of active and free radicals with the substrate, the conversion of the substrate occurs in the secondary reaction, coherently-synchronized with the primary one [3].

The mechanism of such coherently synchronized reactions is being examined in the process of cyclohexane oxidation with hydrogen peroxide in homogeneous and heterogeneous systems. The cyclohexane oxidation is the most commonly studied process, where the valuable for organic synthesis compounds are achieved during oxidative dehydrogenation or during oxidation [4-7]. The implementation of thermodynamically hindered reactions is of great interest for petrochemical and organic synthesis. Among them the reactions of partial dehydrogenation of cyclohexane to cyclohexene and cyclohexadiene should be mentioned.

	ΔG, kJ/mol
$C_6H_{12} \rightarrow C_6H_{10} + H_2$	39.88
$C_6H_{10} \rightarrow C_6H_8 + H_2$	115.37

It is of course more beneficial in terms of energy to carry out such reactions in conjunction with other reactions, in this case, with the reaction of hydrogen peroxide decomposition.

2 Results and discussion

The process of coupling dehydrogenation of cyclohexane to cyclohexene and cyclohexadiene with hydrogen peroxide was carried out without catalyst [8].

The reaction was examined at a temperature range of 450-650 °C at various flow rates and ratios of starting reactants. The process was of sequential and autocatalytic nature with the period of self-acceleration (kinetic curves for cyclohexadiene and benzene), as shown in Figure 1. Under optimum conditions, the yield of cyclohexene was up to 19.4%, of cyclohexadiene was 3.4% and benzene 2.4%.

[a] Corresponding author: tnagiev@azeurotel.com

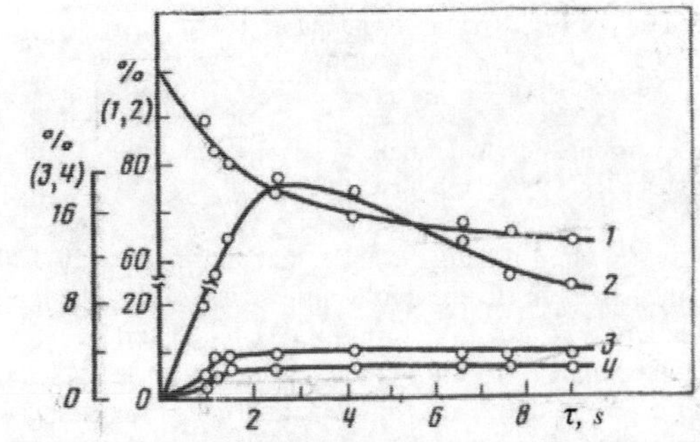

Figure 1. Kinetic curves for coupling dehydrogenation of cyclohexane:

1-cyclohexane; 2-cyclohexene; 3-benzene; 4-cyclohexadiene

Kinetic curves show that the accumulation of cyclohexene as an intermediate product at the initial stage of the process increases up until the rate of its consumption and the rate of accumulation at the conventional contact time equal to τ = 4.2c, becomes equal, and cyclohexene concentration reaches a maximum value. Cyclohexene yield decreases with the further increase in contact time. The rate of formation of cyclohexene and cyclohexadiene initially increases, and reaches its maximum at the inflection point, which is typical for sequential reactions.

At a temperature range of 450-500 °C the cyclohexene is formed, with the increase in temperature range between 560 °C and 650 °C dehydrogenation products contain cyclohexadiene and benzene along with cyclohexene. These data also suggest that with further increase in temperature the ring-opening decomposition reaction of the cyclohexane occurs along with the process of the dehydrogenation. The consequence is the formation of gaseous products such as H_2, CO_2, CH_4, C_2H_4...

These experimental data aggregate show that in the studied range of variations by reaction parameters under the best conditions reaction proceeds in the direction of the oxidative dehydrogenation of cyclohexane achieving the yield of desired products C_6H_{10}, C_6H_8 и C_6H_6 – 19,4%; 3.4% and 2.4% respectively.

It is necessary to identify the nature of the interaction of H_2O_2 with the hydrocarbon, resulting with the formation of unsaturated compound under high-temperature oxidation conditions in order to identify the mechanism of dehydrogenation of cyclohexane.

A prerequisite for the coherently synchronized reactions to proceed in the reaction system is its quantitative characteristic defined by the determinant equation [1]:

$$D = \nu(r_1 / r_{ac} + r_2 / r_{ac})^{-1}$$

where r_1 and r_2 – consumption rate of an actor (H_2O_2) in the primary and secondary reactions respectively, r_{ac} - consumption rate of an acceptor (C_6H_{12}), ν - stoichiometric factor (in our case ν = 1). The value of the determinants calculated by using experimental data D≈0.1 indicates the induced nature of dehydrogenation of cyclohexane.

Under the conditions of the high-temperature cyclohexane gas-phase oxidation with hydrogen peroxide, the reaction proceeds by a free radical mechanism, where active $^{\bullet}OH$ and $HO_2^{\bullet}$ radicals are formed in the primary reaction - in the reaction of H_2O_2 decomposition, and they are consumed in the secondary one - in the dehydrogenation of cyclohexane. Considering that at a high temperature the concentration of $HO_2^{\bullet}$ radicals in the reaction system is much higher than the concentration of $^{\bullet}OH$ radicals [9], the $HO_2^{\bullet}$ radicals play a key role in the mechanism of the dehydrogenation:

chain initiation

$$H_2O_2 \rightarrow 2^{\bullet}OH - 217{,}9 \text{kJ/mol}$$

chain propagation

$$\begin{cases} H_2O_2 + {}^{\bullet}OH \rightarrow HO_2^{\bullet} + H_2O + 107{,}7 \text{kJ/mol} \\ C_6H_{12} + HO_2^{\bullet} \rightarrow C_6H_{10} + H_2O + {}^{\bullet}OH + 40{,}17 \text{kJ/mol} \\ C_6H_{10} + HO_2^{\bullet} \rightarrow C_6H_8 + H_2O + {}^{\bullet}OH + 104{,}6 \text{kJ/mol} \\ C_6H_8 + HO_2^{\bullet} \rightarrow C_6H_6 + H_2O + {}^{\bullet}OH + 66{,}9 \text{kJ/mol} \end{cases}$$

chain termination

$$\begin{cases} {}^{\bullet}OH + wall \rightarrow {}^{\bullet}OH(ads) \\ HO_2^{\bullet} + wall \rightarrow HO_2^{\bullet}(ads) \end{cases}$$

Each successive reaction of oxidative dehydrogenation of cyclohexane is a combination of chain initiation, propagation and termination stages. The system of kinetic equations, adequately describing the experimental data, is as follows [10]:

$$-d[C_6H_{12}]/d\tau = k_{eff1}[C_6H_{12}][H_2O_2]$$

$$-d[C_6H_{10}]/d\tau = k_{eff1}[C_6H_{12}][H_2O_2] - k_{eff2}[C_6H_{10}][H_2O_2]$$

$$-d[C_6H_8]/d\tau = k_{eff2}[C_6H_{10}][H_2O_2] - k_{eff3}[C_6H_8][H_2O_2]$$

where

$$k_{eff1} = 10^{24{,}9} \exp(-264000/RT) sm^3/mol \cdot s$$

$$k_{eff2} = 10^{23{,}9} \exp(-222000/RT) sm^3/mol \cdot s$$

$$k_{eff3} = 10^{22{,}8} \exp(-205000/RT) sm^3/mol \cdot s$$

Examination of the process of coherently synchronized cyclohexane oxidation with hydrogen peroxide was carried out also at a fairly low temperature in a heterogeneous catalytic system. Biomimetic catalyst [11], which simulates the basic functions of the enzyme group of oxidoreductase - catalase and monooxygenase, was used as the catalyst here. Characterised by its highly active and selective action these biomimetic catalysts are synthesized based on iron-porphyrin complexes simulating the active component of cytochrome P-450 [12].

The process of cyclohexane gas phase oxidation was carried out with heterogeneous biomimetic catalysts per-FTPhPFe(III)/Al_2O_3 and PPFe(III)/Al_2O_3 at a temperature of 130 – 250 °C and a molar ratio C_6H_{12}: H_2O_2 = 1:1.7 with 25% aqueous H_2O_2 solution. The results of these studies are shown in Figure 2.

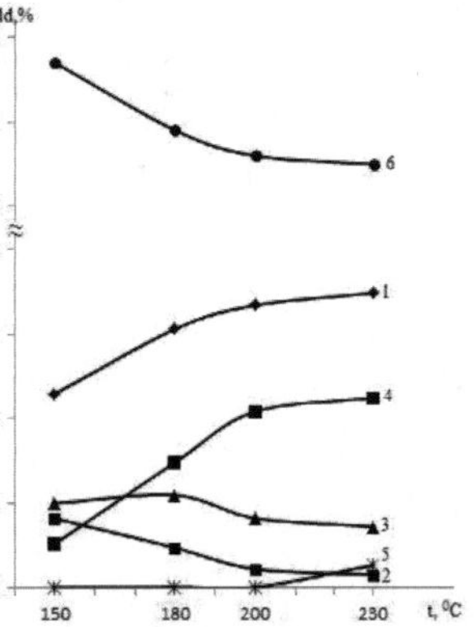

Figure 2. The dependence of the reaction yield of cyclohexane coherently synchronized oxidation with bio simulator per-FTPhPFe(III)/Al_2O_3 on the temperature: [H_2O_2]=25%, $V_{H_2O_2}$ =1,41mL/h, $V_{C_6H_{12}}$ =0,9 mL/h, C_6H_{12}:H_2O_2=1:1,7

1- conversion; 2- cyclohexanol; 3- cyclohexanone; 4- cyclohexene; 5- cyclohexadiene; 6- O_2 yield

As shown in Figure 2 at the temperature up to150-160 °C the reaction that mainly proceeds is cyclohexane oxidation reaction, forming cyclohexanol and cyclohexanone. At the temperature above 180°C the reaction of oxidative dehydrogenation of cyclohexane to cyclohexene (curve 4, Figure 2) begins to accelerate, respectively, the yield of cyclohexanol and cyclohexanone decreases. The sharp increase in the yield of cyclohexene along with the decrease in the yield of formed cyclohexanol indicates that at the temperature of 230 °C or higher cyclohexanol can also be converted into cyclohexene at lower concentrations of H_2O_2 in the reaction system. This conclusion is confirmed by the results of experimental studies of the process of cyclohexanol oxidation with hydrogen peroxide at a lower concentration (20%) with the biomimetic catalyst, where at elevated temperature (200-230 °C) the cyclohexene yield is high (16-30%).

Table 1. The dependence of the reaction yield of cyclohexanol coherently synchronized oxidation with bio simulator per-FTPhPFe(III)/Al_2O_3 on the temperature: [H_2O_2]=20%,

$V_{H_2O_2}$ =1,41 mL/h, $V_{C_6H_{12}}$ =0,9 mL/h, $C_6H_{11}OH$:H_2O_2=1:1

	Composition of feedstock, %				Composition of obtained reaction yield, %						
t, ^{0}C	$C_6H_{11}OH$	2-methyl cyclohexanol	Hexanoic acid	Hexanoic acid cyclohexyl ester	$C_6H_{11}OH$	C_6H_{10}	1,2 cyclo hexane diol	1,3 cyclohexadiene	2-methyl cyclohexanol	other oxygenates	Conversion
150	97,524	0,481	0,712	0,845	83,921	2,321	2,248	0,105	0,093	9,405	14,079
180	97,524	0,481	0,712	0,845	78,186	7,841	2,875	0,145	0,583	10,753	19,814
200	97,524	0,481	0,712	0,845	67,494	16,688	2,887	0,279	0,295	10,652	30,506
230	97,524	0,481	0,712	0,845	55,255	30,374	2,0	0,1	0,641	10,271	42,745

The experiments on the effect of the H_2O_2 concentration at a constant temperature (200 °C) and contact time were conducted. As shown in Figure 3, increased H_2O_2 concentration in the reaction zone contributes to the increase of oxidation reaction rate, which results with the increase of cyclohexanone yield and decrease of cyclohexene yield.

Such results are obtained in a pilot study of cyclohexane oxidation over biomimetic depending on the C_6H_{12} : H_2O_2 ratio at different temperatures.

The results of the effect of the C_6H_{12} : H_2O_2 ratio on the cyclohexane oxidation reaction at a moderate temperature (t = 150 °C), where cyclohexanol and cyclohexanone are the main reaction products are shown in Figure 3.

As it can be seen from the Figure 4 the reaction rate of oxidative dehydrogenation is higher than oxidation reaction rate and the yield of cyclohexane increases at the highest temperature (t = 230 °C).

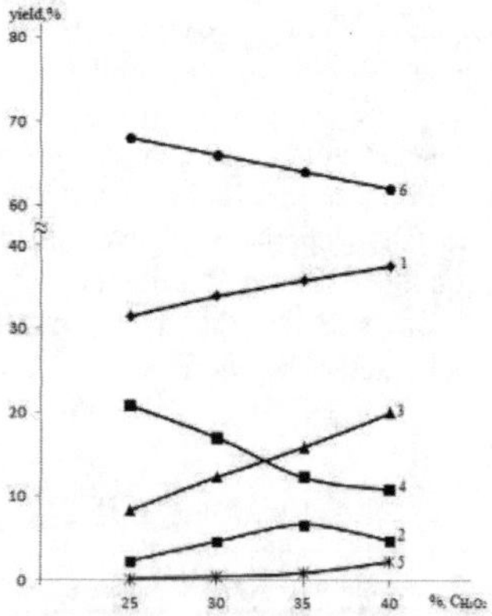

Figure 3. The dependence of the reaction yield of cyclohexane coherently synchronized oxidation with bio simulator per-FTPhPFe(III)/Al_2O_3 on concentration H_2O_2: t=200^{0}C, $V_{H_2O_2}$ =1,41 mL/h,, $V_{C_6H_{12}}$ =0,9 mL/h,,

1- conversion; 2- cyclohexanol; 3- cyclohexanone; 4- cyclohexene; 5- cyclohexadiene; 6- O_2 yield

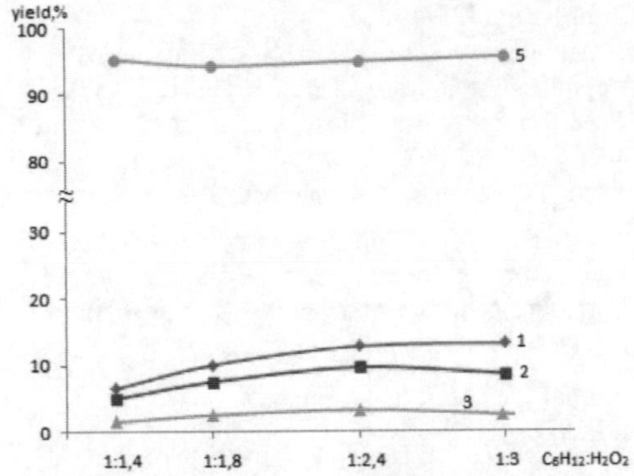

Figure 4. The dependence of the reaction yield of cyclohexane coherently synchronized oxidation over per-FTPhPFe(III)/Al_2O_3catalyst on C_6H_{12} : H_2O_2ratio at the temperature t=150^{0}C,

$V_{H_2O_2}$ =1,41mL/h,, $V_{C_6H_{12}}$ =0,9mL/h,,

1- conversion; 2- cyclohexanol;
3- cyclohexanone; 4- O2yield

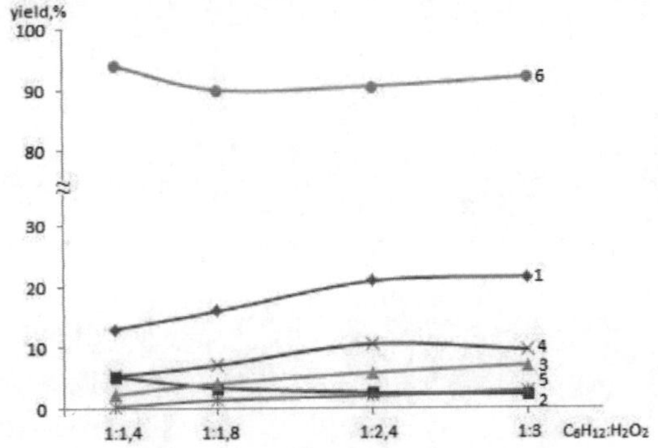

Figure 5. The dependence of the reaction yield of cyclohexane coherently synchronized oxidation over per-FTPhPFe(III)/Al_2O_3catalyst on C_6H_{12} : H_2O_2ratio at the temperature t=230^{0}C,

$V_{H_2O_2}$ =1,41mL/h,, $V_{C_6H_{12}}$ =0,9mL/h,,

1- conversion; 2- cyclohexanol; 3- cyclohexanone; 4- cyclohexene; 5- cyclohexadiene; 6- O_2 yield

Thus, the complex reaction, consisting of parallel-sequential oxidation and dehydrogenation reactions, which are coherently synchronized, proceeds during the process of cyclohexane oxidation with biomimetic catalyst. Depending on the reaction parameters it is possible to deliberately adjust the direction of oxidation reaction and reaction rate.

It should be noted that in terms of the mechanism of each stage of this complex reaction of cyclohexane parallel-sequential oxidation with hydrogen peroxide with biomimetics, each oxidation reaction is carried out under inducing effect of hydrogen peroxide intermediates and consists of two coherently synchronized reactions - 1) catalase (H_2O_2 decomposition reaction) and 2) monooxygenase (cyclohexane oxidation reaction) with biomimetic catalyst.

The probable mechanism of cyclohexane coherently synchronized oxidation, considering the parallel-sequential nature of oxidation and dehydrogenation, can be expressed by the general scheme [3]:

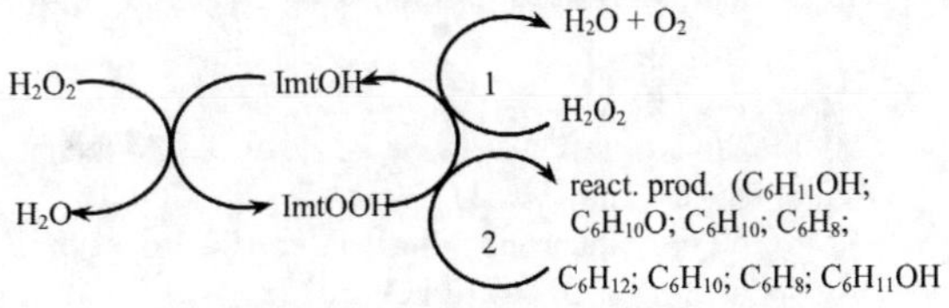

The mechanism of primary and secondary reactions proceeding over biomimetic catalyst is given:

1) Mechanism of catalase reaction

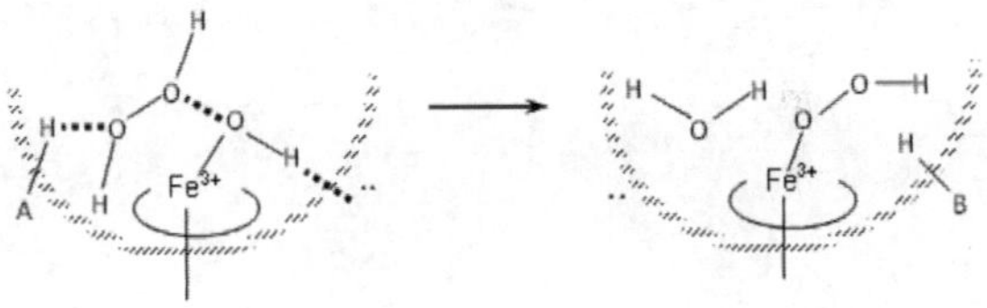

2) The mechanism of reaction of cyclohexane oxidation to cyclohexanol

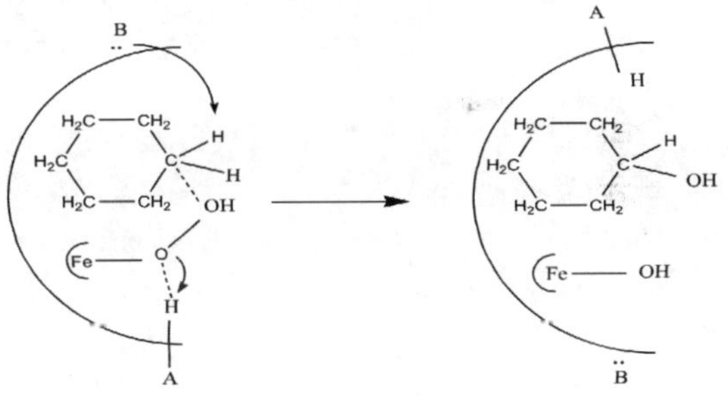

3 Conclusion

The scheme shows that formed active intermediate complex of hydroperoxide (ImtOOH), as in the case of free radical $HO_2^{\bullet}$, is the key active centre for coherent synchronization of catalase (primary) and monooxygenase (secondary) reactions. Coherently synchronized nature of the cyclohexane oxidation reaction follows from the experimental data (Figure 2 and 3). The changes in the catalase reaction rate (curve 6) and the rate of cyclohexane oxidation consumption (curve 1) are shown by the kinetic curves of Figures 2 and 3.

References

[1] T. M. Nagiev, Chemical Conjugation, M.: Nauka, 216 (1989)

[2] T. M. Nagiev, The Theory of Coherent Synchronized Reactions: Chemical Interference Logics, International Journal of Chemical Engineering and Applications, **6**, 5 (2015)

[3] T. M. Nagiev, Coherent Synchronized Oxidation by Hydrogen Peroxide, Elsevier, Amsterdam, 325 (2007)

[4] V. I. Timokhin, M. T. Lisovska, A. P. Pokucha, Kinetics and Catalysis, **41**, 2, pp.179-185 (2000)

[5] Eric C. Tyo, Chunrong Yin, Marcel Di Vece, etc. ACS Catal., **2**, 11, 2409–2423 (2012)

[6] Olsen M. H. N., Salomao G. C., Drago V. J. Supercrit. Fluids, **34**, 2, 119-124 (2005)

[7] Yu K. M. Kerry, Hummeida Ranya, Abutaki Afaf. Catal.Lett., **111**, 1-2, 51-55 (2006)

[8] T. M. Nagiev, F. G. Bayramov, Petrochemistry, **27**, 2, 170-176 (1987)

[9] T. M. Nagiev, S. I. Agaeva, F. G. Bayramov, Z. M. Nagieva, Azerb. Jour. Chem., 4, 16-21 (1997)

[10] T. M. Nagiev, Vestn. Mosc. Univ. Ser. 2. Chemistry, **34**, 3, 260-266 (1993)

[11] T. M. Nagiev, L. M. Gasanova, S. Z. Zulfugarova, etc. Chemical Engineering Communications, An International Journal for Communication of Research. **190**, 5-8, 726-748 (May-August 2003)

[12] U. V. Nasirova, I. T. Nagieva, T. M. Nagiev, Journal of Materials Sciences and Engineering B2(4), 306-312 (2012)

MATEC Web of Conferences **49**, 01003 (2016)
DOI: 10.1051/matecconf/20164901003

Effect of Surfactant Concentration in the Emulsions on the Process of Oleophilic Porous Structures Imbibition

Olga S. Shtyka [a], Jerzy P. Sęk

Lodz University of Technology, Poland

Abstract. The spontaneous imbibition has been a subject of the scientific interest being a background process for numerous industrial technologies and occurring in the natural environment. In literature the experimental and theoretical results regarding this phenomenon describe a media imbibition with single-phase liquids and the relation between the process rate and media characteristics. The imbibition of oleophilic porous structures with two-phase liquids, only one phase of which was wetting, is an objective of the current publication. The main purpose is to estimate the influence of both surfactant fraction and the dispersed phase concentration on the mentioned process. The imbibition rate was investigated during model experiments with stabilized oil-in-water emulsions having the dispersed phase concentrations of 10 vol%, 30 vol% and 50 vol%. The prepared emulsions differed with fraction of the added surfactant, i.e. 1 vol%, 2 vol% and 5 vol%. The obtained results allowed to conclude that at the h_{im}≥0.02 m, the dispersed phase concentration and viscosity decreased versus height. However, the raise of the surfactant fraction caused the increase of mass and height of the imbibed emulsions in porous medium. Moreover, this provided increasing of viscosity and a change of emulsions behaviour as a liquid.

1 Introduction

The spontaneous imbibition is one type of liquid transport in porous structures which is driven by capillary pressure. It appears on pores walls during their wetting by a penetrating liquid, if an adhesion force of liquid molecules exceeds the force of their mutual attraction [1–4]. The imbibition is known as a phenomenon occurring in the natural environment, i.e. hydrological regime in a soil matrix, migration processes in geology; plants seeds imbibition with water. Nowadays, it is applied in numerous industrial spheres: in the environmental engineering (*viz.* oil and its derivatives recovery and absorption); in material engineering (*viz.* percolation of the antifungal emulsions into concrete constructions), as well as in pharmacology and medicine.

The experimental data on the spontaneous imbibition presented in literature can be divided into the following groups: study of the dependence between the process rate (i.e. mass/height change) and time [1, 5–10]; study of the relation between the characteristics of porous structure and the imbibition rate [1, 7–9]; and study of the effect of the penetrating liquid characteristics on this process [5, 6]. Majority of the represented results has been related to the imbibition with single-phase liquids [1, 2, 3, 5–10], while the issues concerning multiphase liquids, *viz.* emulsions, still have not been fully understood and described. It relates with the complexity of the mentioned process.

Emulsions being thermodynamically metastable, in practice can be stabilized by surfactants or solid particles [11–14], or by means of decreasing the droplets size of the dispersed phase [15–17]. However, the addition of stabilization agents is currently the common method used for emulsion stabilization. This explains a great deal of the results regarding the influence of emulsifiers on emulsions stability, size distribution of droplets, their shape, and dispersity [11–14, 18, 19].

The purpose of our research work is to investigate and describe the process of porous media imbibition with two-phase liquids. Concretely, this paper provides the experimental results about the influence of the emulsifier agent fraction and the dispersed phase concentrations on the imbibition process. The estimated values were change of mass and height of the imbibed emulsions as a function of time and variation of emulsions concentration versus height of their front during the imbibition process. In addition to the above, the correlation between the surfactant fraction and the viscosity of the prepared emulsions was investigated.

2 Experimental part

2.1 Materials

In these experiments, stabilized oil-in-water emulsions were used as wicking liquids, which dispersed phase behaved as a wetting one, while its continuous phase was absolutely non-wetting liquid. The studied emulsions had different dispersed phase concentrations φ_d, i.e. 10 vol%,

[a] Corresponding author: olga.shtyka@edu.p.lodz.pl

30 vol%, and 50 vol%. The oily component was represented by the refined vegetable oil produced by EOL Polska Sp.z.o.o., Poland. The viscosity of pure oil was 56.3±0.46 mPa·s and its density was equal to 954.0±1.7 kg/m^3. The surface tension was 32.2±1.3 mN/m. Ethoxylated fatty acid was used as non-ionic surfactant in fractions φ_s of 1 vol%, 2 vol%, and 5 vol%. The commercial name of the emulsifier is Rokacet O7 and it is produced by PCC Exol SA, Poland.

The porous medium was represented in a form of stripes with a height *h* of 0.016 m and a width *a* of 0.035 m. They were cut from a nonwoven polypropylene sheet. The average porosity of investigated medium was equal to 0.93 and the mean thickness was 1.9×10^{-3} m. This material belonged to the type of oleophilic sorbent. According to experimental data, the sorption capacity of vegetable oil (η ~53.1 mPa·s) equaled 24 g/g, while pure water was not wicked. The investigated polypropylene sorbent was obtained from Sintac©–Polska Sp.z o.o., Poland.

2.2 Procedure of the experiment

The dependence between the concentration of the added surfactant and the rate of the imbibition process, changes of the dispersed phase concentration and viscosity of imbibed emulsion versus height were investigated experimentally. Each of the prepared polypropylene stripes was dipped into a beaker with 200 ml of the emulsion [20]. The contact square of material surface was 1.4×10^{-4} m^2.

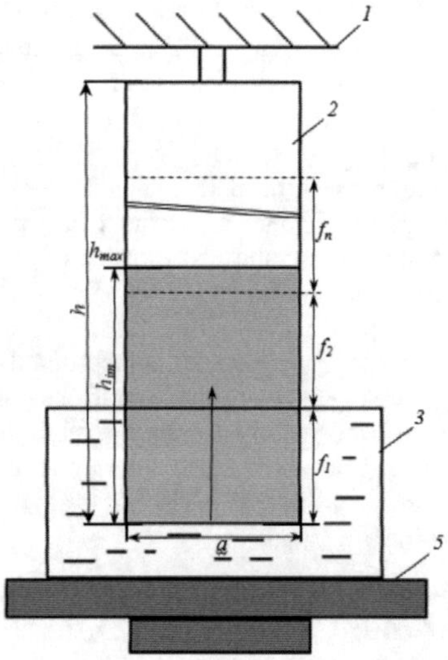

Figure 1. Scheme of the used equipment: 1 – fixator of a porous material; 2 – stripe of a porous material; 3 – reservoir with a liquid; 5 – balance.

The changes of the height of an imbibed liquid front, i.e. h_{im}, and mass m_{im} versus time t_{im} were followed till the time when their values became unalterable, and it was denoted as t_{max}. After that, a sorbent stripe was removed from the beaker and its soaked part was cut into fragments f_n. The size of each fragment was of a width of 0.035 m and a height of 0.02 m (Figure 1). Each sample was put into a ceramic dish and after that, the imbibed emulsion was washed out with distilled water.

In obtained mixture the concentration of the dispersed phase was measured by means of the nephelometrical method using of the optical analyzer TurbiscanTM LAB (Formulaction, France). This method has been reported elsewhere [21].

2.3 Analytical methods

The density of the prepared emulsions was measured by means of the picnometric method. The viscosity changes related to increasing of the surfactant and inner phase concentrations were determined using a rotational viscometer Rheometer RC 20 (Rheotec, Germany). The shear rate values varied and were in a range of 80–140 1/s. The surface tension was measured with a tensiometer KRÜSS K100, Germany.

All experiments were conducted at a temperature of 23±1°C and atmospheric pressure.

3 Results and discussion

3.1 Viscosity changes vs surfactant concentration increasing

According to the obtained data, the increase of a surfactant concentration caused changes of viscosity of the investigated emulsion and their behavior as a liquid. As shown in Figure 2, the flow curves can be approximated by a straight line in case of all emulsions with a dispersed phase concentration of 10% and 30% emulsions with $\varphi_s \leq 2$ vol%. Thus, it means that the investigated disperses behaved as Newtonian liquids, while other emulsions, mainly with higher concentrations of inner phase, behaved as non-Newtonian liquids.

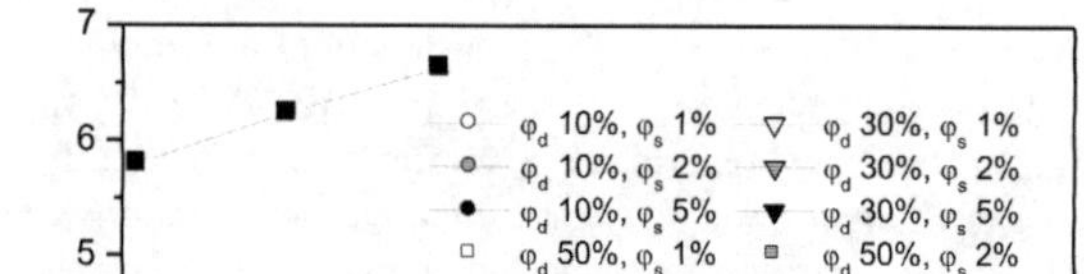

Figure 2. Dependence between shear stress and the surfactant concentration at different shear rates.

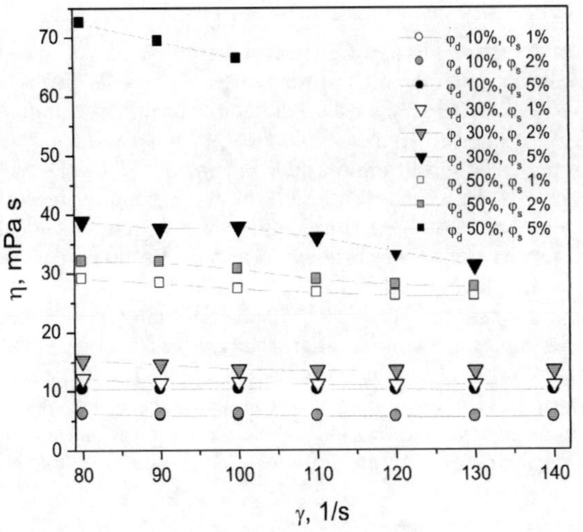

Figure 3. Changes of viscosity versus increasing of the surfactant concentration at different shear rate.

The same conclusions were made based on viscosity data analyzing. Its values were constant at different shear rate for all 10% emulsion and 30% emulsions with φ_s =1 and 2 vol%. Others behaved as the non-Newtonian shear thinning liquids, thus, increasing of the shear rate caused decreasing of the viscosity value. For 30% emulsion it reduced from 38.9 mPa·s for γ=80 1/s till 31.2 mPa·s for γ=130 1/s and for another one, viscosity was 72.5 mPa·s at shear rate 80 1/s, while for 1000 1/s, it decreased to 66.4 mPa·s (Figure 3).

3.2 Changes of imbibition rate vs increasing of surfactant concentration

The preliminary experiments showed that the process of porous media imbibition with emulsions depended on a fraction of added surfactant and this revealed to be a variation of the height and weight of imbibed liquids. The results concerning obtained maximal mass of the liquids imbibed in the non-submerged part of sorptive media, are presented in Table 1.

Table 1. Maximal values of mass, [kg].

Fraction of surfactant, φ_s	The dispersed phase concentration, φ_d		
	10 vol%	30 vol%	50 vol%
1 vol%	0.0015	0.0009	0.0004
2 vol%	0.0016	0.0012	0.0011
5 vol%	0.0024	0.0022	0.0012

The increase of emulsifier fraction caused a raise of the imbibed emulsion mass in all investigated cases. For 50% emulsions, the value of imbibed mass was 3 times higher at φ_s=5 vol% comparing to the emulsion with the lowest surfactant fraction, i.e. 1 vol%. Such ratio was equaled to 2.5 for 30% emulsion and 1.6 for 10%, respectively. The same tendency was observed for the changes of height of the imbibed emulsions. As an example, the results for 10% and 50 % emulsions are plotted in Figure 4–5.

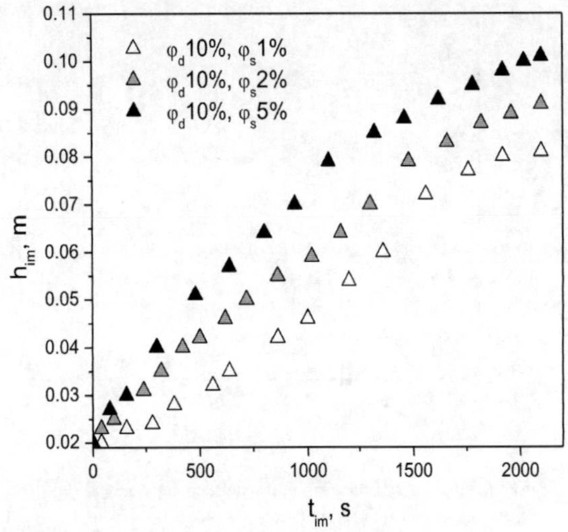

Figure 4. Change of an imbibed liquid height vs time for emulsions with different fractions of surfactant (φ_d =10 vol%).

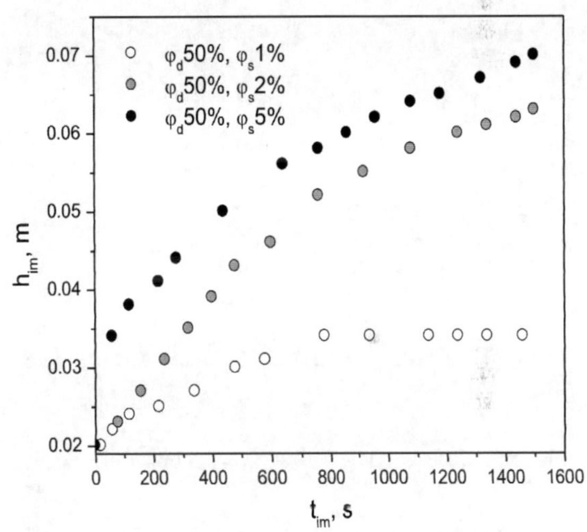

Figure 5. Change of an imbibed liquid height vs time for emulsions with different fractions of surfactant (φ_d =50 vol%).

In case of 10% emulsions, it increased from 0.081 m (φ_s=1 vol%) to 0.101 m (φ_s=5 vol%), i.e. by 25 %. For 30% emulsions, the difference of the fronts heights was 0.02 m (~30%). Whereas, in case of 50% emulsions, the increasing of surfactant fraction till 5 vol% caused rising of the height h_{im} by 75%, *viz.* from 0.04 to 0.07 m.

3.3 Changes of the dispersed phase concentrations vs height

The changes of the dispersed phase concentrations versus the height of an imbibed emulsion are shown in Figure 6–8. On the graphs, the vertical axis represents the normalized values of the dispersed phase concentrations which were calculated as a ratio φ_{im}/φ_d where φ_{im} is the obtained experimentally value of the concentration in an imbibed emulsion and correspondingly, φ_d is defined as the initial concentration of the dispersed phase.

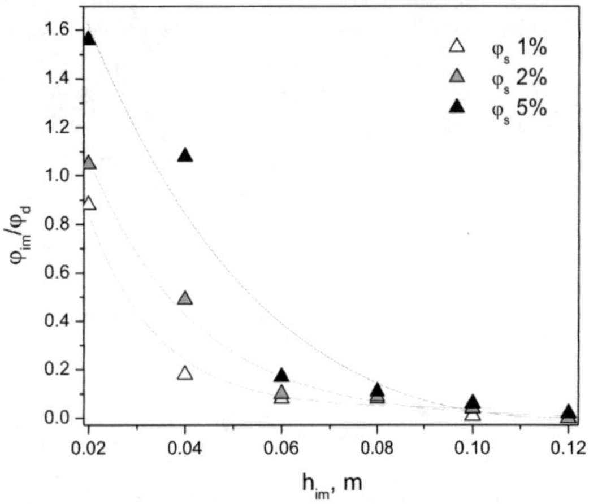

Figure 6. Changes of the emulsions concentrations vs height for different fractions of surfactant (φ_d =10 vol%).

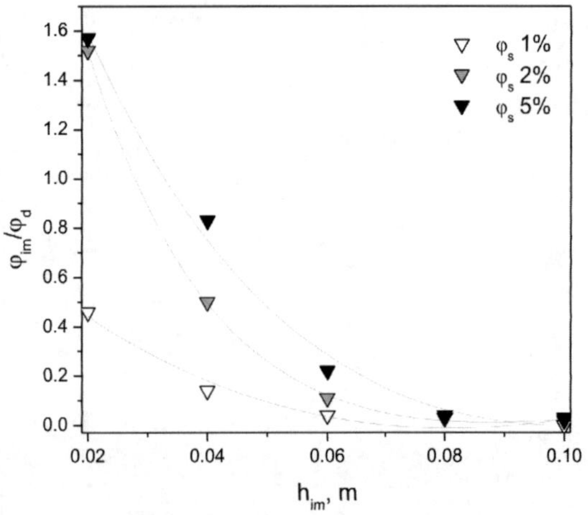

Figure 7. Changes of the emulsions concentration vs height for different fractions of surfactant (φ_d =30 vol%).

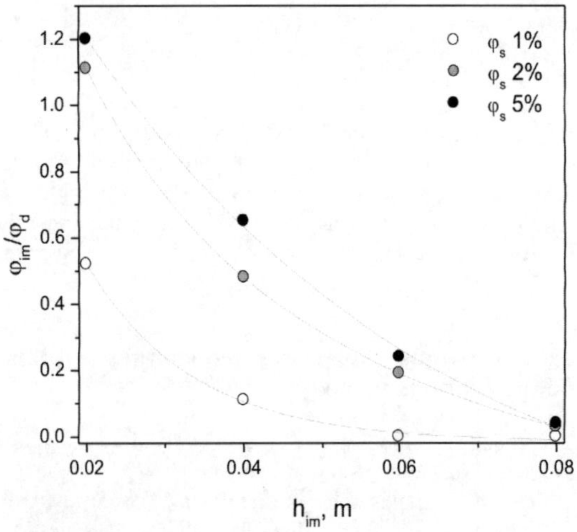

Figure 8. Changes of the emulsions concentration vs height for different fractions of surfactant (φ_d =50 vol%).

In all investigated cases, the obtained concentrations of dispersed phase were lower in case of φ_s=1 vol%, and consequently, the highest were observed for φ_s=5 vol%.

At the height h_{im} ≤0.02 m, the obtained concentration of the imbibed emulsion exceeded its initial value φ_d, but only when surfactant was added in amount of 2 vol% and 5 vol%. However, at h_{im}=0.08 m, the obtained values of the dispersed phase concentration were almost identical for emulsions with different surfactant fractions, but the same content of oil phase (Figure 6–8).

In case of 30% and 50% emulsions, the curves describing the concentration changes, beside φ_s=1 vol%, were relatively close to each other (Figure 7–8). It proves that for emulsions with a higher dispersed phase concentration (φ_d ≥30 vol%), the increasing of surfactant concentration caused little effect on the investigated process.

3.4 Changes of the emulsion viscosity vs height

As previously mentioned, the influence of added surfactant was estimated from the point of their stabilization and the decrease rate of deemulgation process. However, the surfactants influence also on the emulsion ability to penetrate through porous structure.

The obtained results concerning changes of the emulsion viscosity versus height are shown in Figure 9–11.

In case of surfactant concentration of 1 vol%, the tendency of the viscosity decreasing was observed for all investigated emulsions (Figure 9). At h_{im}≤0.02 m, the obtained values of viscosity were approximately the same as the initial ones. For emulsion with φ_d=10 vol%, the front of imbibed liquid was higher, but the viscosity was the lowest. However, the maximal value of viscosity was observed for 50% emulsion at h_{im}≤0.04 m.

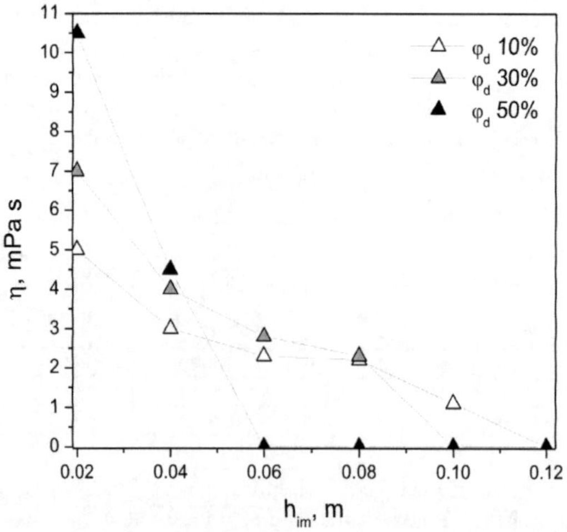

Figure 9. Changes of the viscosity vs height for different values of the dispersed phase concentration (φ_s=1 vol%).

As shown in Figure 10, at h_{im}≤0.02 m, the obtained viscosity values were a few higher in comparison with initial data for all emulsions with φ_s=2 vol%. However, there was a tendency of the viscosity decrease at $h_{im}\geq 0.02$ m.

As in the previous case, the maximal values of viscosity were obtained for the emulsion with φ_d =50 vol%, but for $h_{im}\leq 0.08$ m, consequently the lowest were observed for 10% at the height less than 0.10 m (Figure 10).

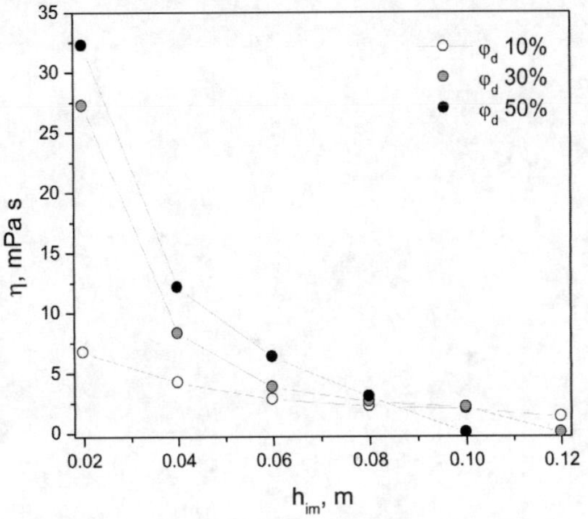

Figure 10. Changes of the viscosity vs height for different values of the dispersed phase concentration (φ_s=2 vol%).

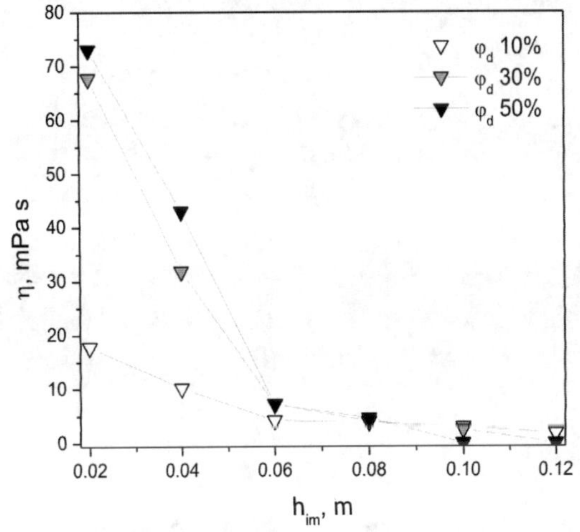

Figure 11. Changes of the viscosity vs height for different values of the dispersed phase concentration (φ_s=5 vol%)

The effect of the surfactant addition in an amount of 5 vol% was exceeding of viscosity values obtained for the initial emulsions. This phenomenon was observed at height $h_{im}\leq 0.02$ m and the main reason can be the increase of the phase concentration as shown in Figure 6–8.

However, at $h_{im}\geq 0.02$ m, the viscosity of an imbibed emulsion decreased with increasing of their front height and it was observed for all investigated emulsions.

4 Conclusions

To sum up, the amount of the surfactant influenced the rheological properties of the investigated liquids.

For example, the emulsions with a low concentrated dispersed phase, i.e. φ_d<30 vol%, and surfactant volume fraction of φ_s<2 vol% behaved as Newtonian liquids, while at $\varphi_d\geq 30$ vol% and $\varphi_s\geq 2$ vol%, they exhibited properties of non-Newtonian liquids.

Moreover, the process of porous structures imbibition with surfactant-stabilized emulsions depended on both the initial concentration of the inner phase and the surfactant fraction. Firstly, it revealed increasing of the weight and height of the imbibed emulsions with a raise of surfactant fraction.

In contrast, their decreasing was observed simultaneously with an increase of the dispersed phase concentration. The obtained results also allowed to conclude that at $h_{im}\geq 0.02$ m, the inner phase concentration and viscosity of imbibed emulsion decreased versus height of its front in a porous medium.

Acknowledgment

The study was funded in framework of the Foundation Program for Young Scientists in Lodz University of Technology (Grant No. 501/10-34-2-1015).

References

1. Masoodi R., Pillai K.M., AIChE, **56**, 9 (2010)
2. Alava M., Dube M., Rost M., Adv. Phys. **53** (2004)
3. Zhmud B.V., Tiberg F., Hallstensson K. J., Coll. Interf. Sci., **228** (2000)
4. Chatterjee P. K., Gupta B. S. Textile Science and Technology, **13** (2002)
5. Xue H.T., Fang Z.N., Yang Y., Huang J.P., Zhou L.W., Chem. Phys. Lett., **432** (2006)
6. Siebold A., Nardin M., Schultz J., Walliser A., Oppliger M., Coll. Surf., **161** (2000)
7. Hamraoui A., Nylander T., J. Coll. Interf. Sci., **250**, 2 (2002)
8. Masoodi R., Pillai K.M., Varanasi P.P., AIChE J. **53**, 11 (2007)
9. Dang-Vu T., Hupka J., Phys. Pr. Miner. Proc., **39** (2005)
10. Cai J., Hu X., Standnes D. C., You L., Coll. Surf. A: Physicochem. Eng. Aspects, **414** (2012)
11. Blink B.P., Lumson S.O., Langmuir, **16** (2000)
12. Bibette J., Leal-Calderon, Curr. Opin. Colloid Interface Sci., **1** (1996)
13. Blink B.P., Current Opinion Coll. Interf. Sc., **7** (2002)
14. Leal-Calderon F., Schmitt V., Curr. Opin. Colloid Interface Sci., **13** (2008)
15. Rahn-Chiquea K., Puertas A.M., Romero-Canob M.S., Rojasa C., Urbina-Villalbaa G., Adv. Coll. Interf. Sci., **178** (2012)
16. Tadros T., Izguierdo P., Esquena J., Solans C., Adv. Coll. Interf. Sci., **108** (2004)
17. Gutierrez J.M., Gonzalez C., Maestro A., Sole I., Pey C.M., Nolla J., Curr. Opin. Colloid Interface Sci., **13** (2008)
18. Binks B.P., Jhonny A.R., Langmuir, **23**, 7 (2003)

19. Behrend O., Ax K., Schubert H., Ultrasonics Sonochem., **7** (2000)
20. Sęk, J., Shtyka O., Szymczak K., J. Environ. Eng. Landsc., **23**, 2 (2015)
21. Lemarchand C., Couvreur P., Vauthier C., Costantini D., Gref R., Int. J. Pharm., **254** (2003)

MATEC Web of Conferences **49**, 02001 (2016)
DOI: 10.1051/matecconf/20164902001

Synthesis of Functionalized Polythiophene as a Potenial Organic Semi-Conductor

Mohammad Siddiqui, Nisar Ullah, Abdurrahman Al-Betar and Abdulaziz Al-Saadi

Chemistry Department, King Fahd University of Petroleum and Minerals, Dhahran 31261, Saudi Arabia

Abstract. The synthesis Poly[1,5-naphthyridine-(3-hexylthiophene)] (PN_3HTh) semi-conducting polymer has been accomplished by adopting both conventional and microwave-assisted Suzuki-Miyaura cross-coupling reaction between 3-hexylthiophene-2,5-diboronic ester and 2,6-dibromo-1,5-naphthyridine. The electrochemical and transport properties of PN_3HTh were investigated both in the bulk as well as in thin film form. These properties can be further tuned by changing the solvent and the nature of the electrode used. Consequently, cyclic voltammetry (CV) measurements were conducted using a potentiostat coupled with an electrochemical cell. The CV results of PN_3HTh as a bulk form in H_2SO_4 solution indicated that the concentration of the polymer in the solution is not well defined because of polymer poor solubility in aqueous solutions. On the other hand, the sulfur group which works as electron donating makes the system more electron-rich. This can explain by the absence of the reduction peak. For the thin film, two single oxidation peaks were obtained at around 0 V and 0.3 V for both cases. Different solvents can tune the transport properties of the polymer as can be seen from the two CVs where BF3 exhibited enhanced transport properties over ACN. The synthesized polymers were characterized by modern spectroscopic methods including IR and NMR.

1 Introduction

During recent years, conjugated polymers and oligomers have been under intensive research and development as new functional materials for electronics [1-5]. These electronic polymers, inherently (or intrinsically) conductive polymers (ICPs) contain an extended π-conjugated organic backbone, rendering unique opto-electrical properties. The ICPs possess the electrical properties of metals or semiconductors while exhibiting the mechanical properties and processibilites of polymers. Applications of ICPs include, photonics, advanced coatings, electromagnetic-interference (EMI) shielding, light-emitting diodes (OLEDs, organic field effect-transistors (OFETs), and anti-corrosion coatings for iron and steel. Several classes of polyconjugated molecules of high molecular weight (polymers), and of relatively low molecular weight (oligomers), have been synthesized. Polythiophenes, polypyroles and other polymers with heteroatoms in the polymer main chain are considered among the most important classes of heterocyclic polymers with extended n-conjugation [6-10]. The preparation of such polymeric materials often involves metal-catalyzed cross-coupling reactions, and frequently requires long reaction time to yield high-molecular-weight material.

Since the electronic properties of semiconducting polymers are often strongly molecular weight dependent, research to develop high molecular weight regioregular polythiophenes (RPTs) have been a matter of great significance. There is an ever-increasing body of work, which shows that microwave heating is efficient in reducing reaction times and, consequently, increases reaction yields and reduces the amount of side products. The real advantage of microwave radiation is that it is generally ‘quicker and cleaner’ than conventional heating, reducing reaction times by more than a factor of 10 and yielding products in high yield with less side-products.

In this study, the synthesis of poly[1,5-naphthyridine-(3-hexylthiophene)] (PN_3HTh) semi-conducting polymer was accomplished by adopting both conventional and microwave-assisted Suzuki-Miyaura cross-coupling reaction between 3-hexylthiophene-2,5-diboronic ester and 2,6-dibromo-1,5-naphthyridine. The synthesized polymers were characterized by modern spectroscopic methods including IR and NMR. The electrochemical and transport properties of PN_3HTh were investigated both in the bulk as well as in thin film form.

2 Experimental work

Instrumentation: Melting Points were determined on a Büchi apparatus (Büchi Labortechnik AG, Switzerland) and are uncorrected. Elemental analysis was carried out on a Perkin Elmer Elemental Analyzer Series 11 Model 2400 (PerkinElmer Inc. USA). IR spectra were recorded on a Perkin Elmer 16F PC FTIR spectrophotometer (PerkinElmer Inc. USA). 1H and 13C NMR spectra were measured in CDCl3 and d6-DMSO using TMS as internal standard on a JEOL JNM-LA 500 MHz spectrometer (JEOL USA Inc.). Mass spectra were recorded on Agilent Technologies 6890N GC–MS system (Agilent Technologies Inc. USA). Analytical TLC was carried out on silica gel 60 F254 plates (E. Merck); column chromatography was carried out on silica gel (200-400 mesh, E. Merck).

Synthesis of 6-Chloro-2-methyl-3-nitropyridine (2): To a solution of 6-methyl-5-nitropyridin-2-amine (20 g, 130.72 mmol) in water (330 ml) at 0°C was added sulfuric acid (23.6 ml) slowly followed by the dropwise addition of an ice-cold solution of sodium nitrite (12.0 g, 173.91 mmol) in water (30 ml). The resulting solution was allowed to react, with stirring, for 20 at 0°C and then allowed to warm to room temperature and stirred for an additional one hour. The resulting slurry was filtered and the solids collected by vacuum filtration to afford the title compound 6-Chloro-2-methyl-3-nitropyridine **2** as a brown solid solid [11].

Synthesis of 6-Methoxy-2-methyl-3-nitropyridine (3): A roundbottom flask was charged with a solution of 6-chloro-2-methyl-3-nitropyridine (12.0 g, 69.36 mmol) in methanol (100 ml). Sodium methanolate (5.6 g, 103.70 mmol) was added and the resulting solution was stirred at room temperature for 24 hours. The mixture was poured into 100 ml of water and the solids that formed were collected by filtration to afford the title compound as light yellow solid.

Synthesis of ethyl 3-(6-methoxy-3-nitropyridin-2-yl)-2-oxopropanoate (4): A roundbottom flask was charged with ethanol (54 ml). To this was added potassium (8.2 g, 0.21 mol) and mixture was stirred until all the potassium dissolved. Diethyl oxalate (30.6 g, 0.21 mol) was then added. The mixture was stirred for 5 minutes and then 6-methoxy-2-methyl-3-nitropyridine (32.0 g, 0.19 mol) was added. The resulting solution was stirred at room temperature for 36 hours. The slurry was filtered and the collected solids were washed thoroughly with ether and dried under reduced pressure. The solid was then suspended in 30 ml of water acetic acid was added until the pH - 4. The solids that formed were collected by filtration to afford the title compound as a light yellow gum.

Synthesis of 3-hydroxy-6-methoxy-3,4-dihydro-1,5-naphthyridin-2(1H)-one (5): A roundbottom flask was charged with a solution of ethyl 3-(6-methoxy-3-nitropyridin-2-yl)-2-oxopropanoate (3.73 mmol) in a solution of ethyl acetate (50 mL) and methanol (20 mL). The mixture was cooled to 0 °C and a solution of sodium borohydride (0.5 mg, 0.01 mmol) in methanol (30 ml) was added. Aluminum trichloride (50 mg, 0.37 mmol) was then added and the resulting solution was stirred for 4 hours at 0 °C. The reaction mixture was then quenched by the addition of ice water. The resulting solution was diluted with 100 ml of water and extracted two times with 100 ml of ethyl acetate. The combined organic layers were concentrated under reduced pressure to afford the ethyl 2-hydroxy-3-(6-methoxy-3-nitropyridin-2-yl)propanoate as a light yellow solid. A roundbottom flask was charged with a solution of ethyl 2-hydroxy-3-(6-methoxy-3-nitropyridin-2-yl)propanoate (37 mmol) in methanol (500 ml). To the mixture was added 10% palladium on carbon (3 g). The solution was shaken under 40 psi of hydrogen for 12 hours. The resulting solution was then allowed to react, with stirring, for 5 hours while the temperature was maintained at room temperature. The mixture was then filtered and the filtrate was diluted with 100 ml of water. Hydrochloric acid was added until the pH = 5. The mixture was filtered and the filtrate was adjusted to pH = 8 by the addition of sodium carbonate. The solids that formed were collected by filtration to afford the title compound 3-hydroxy-6-methoxy-3,4-dihydro-1,5-naphthyridin-2(1H)-one **5** as a thick gum.

Synthesis of 2,5-Bis(4,4,5,5-tetramethyl-1,3,2-dioxaborolane)-3-hexylthiophene (8): Under nitrogen atmosphere, n-Butyllithium (3.8 mL of a 2.5 M solution in hexane, 9.5 mmol) was slowly added to a solution of 2,5-dibromo-3-hexylthiophene (1.30 g, 4.0 mmol) in THF (30 mL) at -78 °C and the mixture was stirred for 2 h at the same temperature. Afterward, 2-Isopropoxy-4,4,5,5-tetra-methyl-1,3,2-dioxaborolane (2.4 mL, 11.6 mmol) was promptly injected into the reaction mixture and the reaction was stirred for 1 h at -78 °C followed by stirring continued overnight at room temperature. The mixture was poured onto ice-water and extracted with dichloromethane (50 mL). The organic layer was dried over anhydrous magnesium sulfate and concentrated under reduced pressure. Column chromatography of the light orange oily material, using boric acid impregnated silica gel, eluting with acetate/hexane, 1/10, v/v) gave the title compound **8** as a colorless viscous oil (0.84 g, 50% yield). 1H-NMR [500 mHz] ($CDCl_3$, δ/ppm): 7.52 (s, 1H), 2.88 (t, 2H), 1.60 (m, 2H), 1.34 ~1.28 (m, 30H), and 0.88 (t, 3H). 13C-NMR [500 mHz] ($CDCl_3$, δ/ppm): 155.084, 139.592, 84.013, 83.583, 31.768, 31.669, 29.919, 28.978, 24.769, 22.598, and 14.130. FTIR: 3032 cm-1 (=C-H, m), 2922.5 cm-1 (-C-H,s), 1575.3 cm-1 (C=C, m), 1473.8 cm-1 (-CH2, s), 1322.5 cm^{-1} (-CH2, s), 1060 cm-1 (-C-O, m), 697 cm-1 (C-S, s).

Synthesis of Poly[1,5-naphthyridine-(3-hexylthiophene)] (9) by conventional method: Under an argon atmosphere, to a solution of compound **8** (0.25 g, 0.6 mmol), compound **9** (0.17 g, 0.6 mmol) in toluene (18 mL) was sequentially added $Pd(PPh_3)_4$ (70 mg, 0.06 mmol) and 2 M K_2CO_3 (3 mL). After bubbling argon through the solution for 2 minutes, the reaction mixture was at 90 °C for 72 h. After cooling to room temperature, the mixture was extracted with ethyl acetate (20 mL) and the organic layer was washed with water and then dried with Na_2SO_4. Evaporation of votiles under reduced pressure yielded crude powder which was washed with methanol and acetone by Soxhlet extraction for a day. The obtained precipitate was collected to give dark green solids in a yield of (yield 89%). ^{1}H-NMR [500 MHz] ($CDCl_3$, δ/ppm): 7.69 (aromatic, -CH=N), 7.55 (thiophene, -CH=C), 7.49 (aromatic, -CH=C), 2.66 (=C-CH_2-CH_2), 1.69 ((=C-CH_2-CH_2-CH_2), 1.33 ~1.26 (CH_2-CH_2-CH_2-CH_2-CH_2-; -C-$(CH_3)_2$), and 0.90 (-CH_2-CH_3). FTIR: 3037 cm-1(=C-H, m), 2995 cm-1(-C-H, s), 1581 cm-1 (C=C, m), 1467 cm-1 (-CH_2, s), 1387 cm-1 (-CH_2, s), 697.6 cm-1 (C-S, s).

Synthesis of Poly[1,5-naphthyridine-(3-hexylthiophene)] (9) by microwave assisted method: In a microwave tube, to a solution of mixture of compound **8** (0.1 g, 0.24 mmol), **6** (0.07 g, 0.24 mmol) in H_2O (2 mL) was added tetra-n-butylammonium bromide (0.18 g, 0.56 mmol) followed by the addition of Na_2CO_3 (0.06 g, 0.56 mmol). After the nitrogen gas was bubbled through the solution for two minutes, [1,1'-Bis(diphenylphosphino)-ferrocene]palladium(II) dichloride (10 mol %) was added and the mixture was heated for 15 minutes at 80 ^{0}C at. The product was isolated by washing with methanol and decanting to get the desired **9** in 52% yield. ^{1}H-NMR and FTIR of the synthesized material were in full agreement with the previously synthesized material through conventional methods.

Raman Spectroscopy of Poly[1,5-naphthyridine-(3-hexylthio-phene)] (9): Raman spectra were recorded on HORIBA LabRam HR high resolution spectrometer with Ar+ polarized laser green line. To avoid thermal effects of the laser beam on the films, intensity was adjusted to an excitation power of 50 μ W. The acquisition time is 400 s. Spectra fitting was done with LabSpec© software supplied with the apparatus.

Field - Emission Scanning Electron Microscope (FESEM) of Poly[1,5-naphthyridine-(3-hexylthio-phene)] (9): The morphological and detailed microstructural attributes of the materials were discerned with the aid of Field Emission Dual Beam (Electron/ Focused Ion Beam) system, and that combines a high - end field - emission scanning electron microscope (FESEM) and a high - performance focused ion beam (FIB) system in one chamber (Tescan Lyra-3). Electron guns are used to produce a fine, controlled beam of electrons which are then focused at the specimen surface. The electron gun omits electrons from field-emission gun to produce the image and number of other characteristics.

Cyclic Voltammetry of Poly[1,5-naphthyridine-(3-hexylthio-phene)] (9): The electrochemical and transport properties of **9** were investigated both as a bulk form and as a thin film form. Therefore, CV measurements were conducted using a potentiodtat / galvanostat coupled with an electrochemical cell. The supporting electrolyte was a 1.0 mol L^{-1} sulfuric acid solution. The cell was composed of an Ag/AgCl as reference electrode, a glassy carbon electrode (GCE) as working electrode, and a platinum wire as counter electrode. The scan rate was fixed at 50 mV s^{-1}.

3 Results and discussion

The synthesis of polymer **9** necessitated the synthesis of monomer **6** which in turn was synthesized as outlined in scheme I. In short, reaction of pyridine **1** with a mixture of phosphoryl chloride and phosphorous pentachloride rendered the known **2** [11] in high yield. Displacement of the chloro group of **2** with a methoxy group, using sodium methoxide in methanol finished intermediate **3** which was in turn transformed to the desired **4** by the action of treatment with a base followed by condensation of the resultant carbanion with diethyl oxalate. Reduction of the nitro and carbonyl functions of intermediate **4** with aluminium borohydride, generated in situ by the reaction of sodium borohydride and aluminium trichloride, followed by concomitant lactamization furnished **5** in good yield. The aromatization of **5** was ensued by the action of phosphoryl bromide in refluxing DMF to generate the desired **6** in good yield (Scheme I).

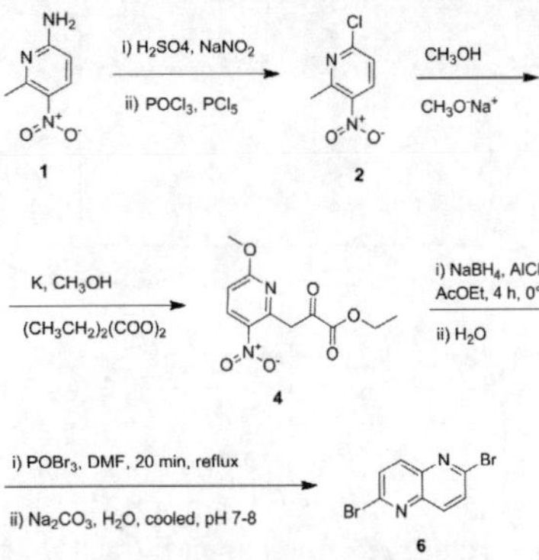

Scheme I. Synthesis of 2,6-dibromo-1,5-naphthyridine (**6**)

The characteristic FTIR, proton and carbon spectra of 2,6-Dibromo-1,5-naphthyridine (6) are given below in Figures 1–3.

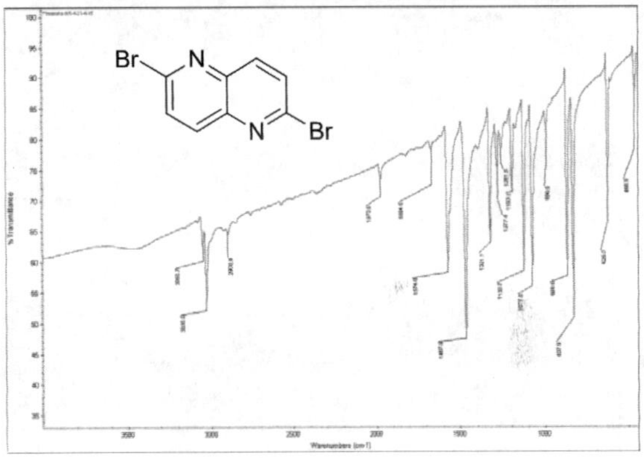

Figure 1. FTIR spectra of 2,6-Dibromo-1,5-naphthyridine (6)

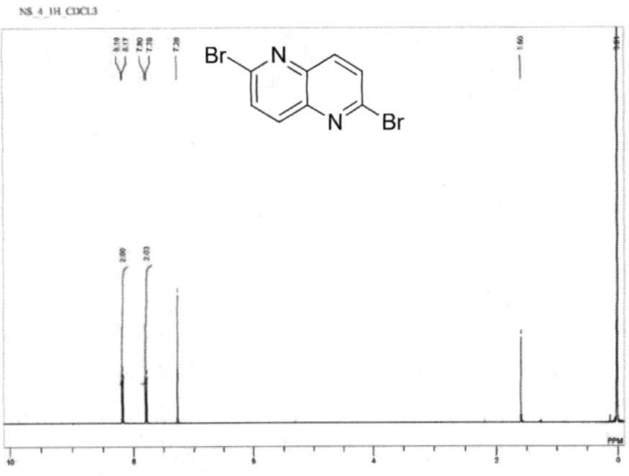

Figure 2. Proton NMR spectra of 2,6-Dibromo-1,5-naphthyridine (6)

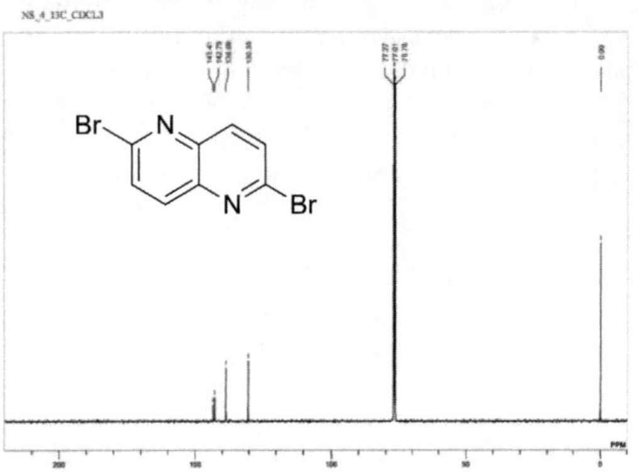

Figure 3. Carbon NMR spectra of 2,6-Dibromo-1,5-naphthyridine (6)

Having secured the synthesis of monomer **6** we next moved towards the synthesis of the monomer **8**. To this end, lithium bromine exchange of compound **7** at lower temperature followed quenching the anion with isopropoxy borolane gave monomer **8** in good yield (Scheme II). The characteristic FTIR, proton and carbon spectra of 2, 5-Dibromo-3-hexylthiophene (7) are given below in Figures 4–6.

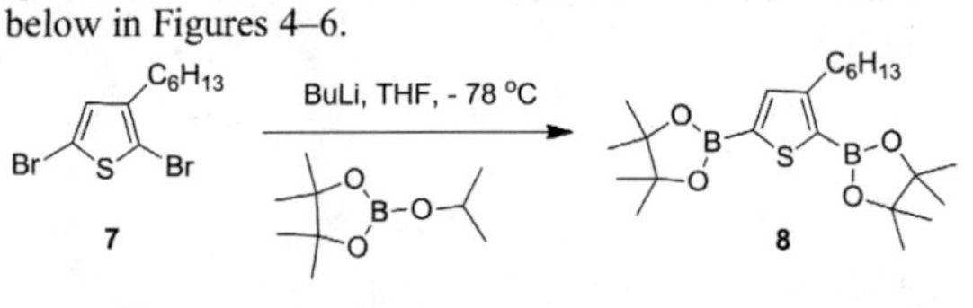

Scheme II. Synthesis of monomer (**8**)

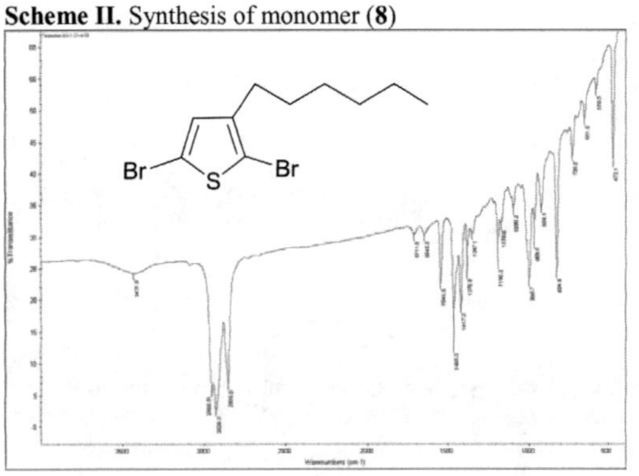

Figure 4. FTIR spectra of 2, 5-Dibromo-3-hexylthiophene (7)

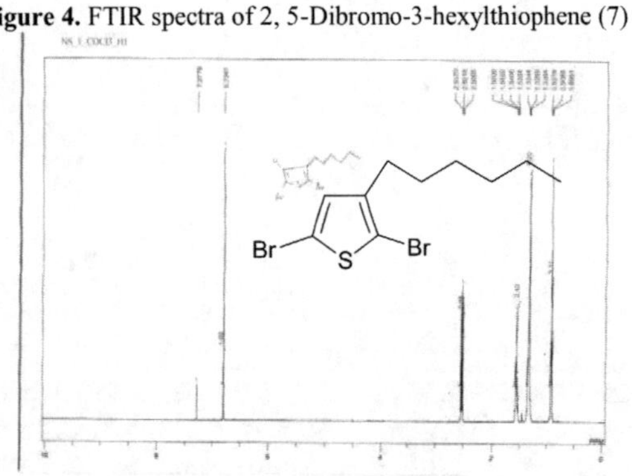

Figure 5. Proton NMR spectra of 2, 5-Dibromo-3-hexylthiophene (7)

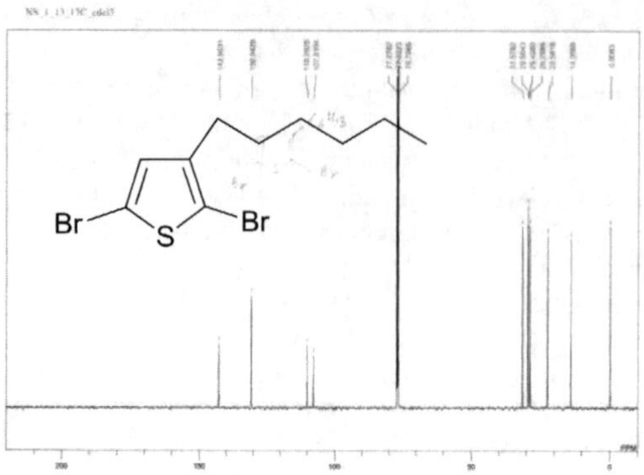

Figure 6. Carbon NMR spectra of 2, 5-Dibromo-3-hexylthiophene (7)

The characteristic FTIR, proton and carbon spectra of 2,5-Bis(4,4,5,5-tetramethyl-1,3,2-dioxaborolane)-3 hexylthiophene (8) are given below in Figures 7–9.

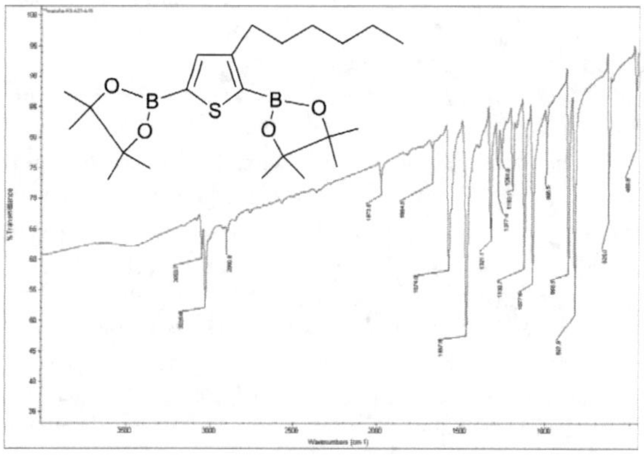

Figure 7. FTIR spectra of 2,5-Bis(4,4,5,5-tetramethyl-1,3,2-dioxaborolane)-3-hexylthiophene (8).

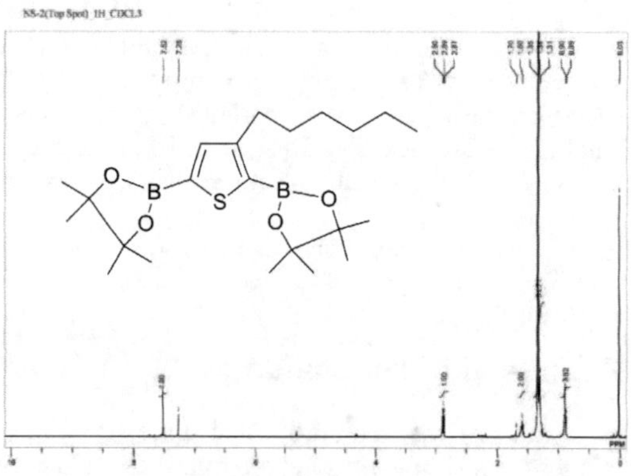

Figure 8. Proton NMR spectra 2,5-Bis(4,4,5,5-tetramethyl-1,3,2-dioxaborolane)-3-hexylthiophene (8).

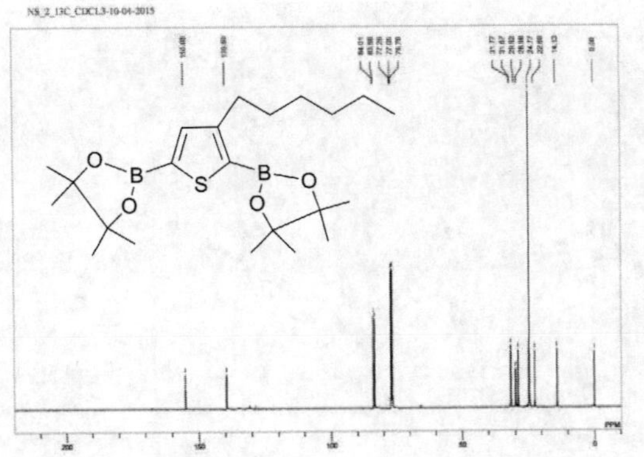

Figure 9. Carbon NMR spectra of 2,5-Bis(4,4,5,5-tetramethyl-1,3,2-dioxaborolane)-3-hexylthiophene (8).

For the polymerization reaction we first tried microwave assisted polymerization by treating the monomers under microwave heating at 80 °C for 20 minutes, using 1,1'-bis(diphenylphosphino)ferrocene]-palladium(II) dichloride as a catalyst. Unfortunately this reaction was very sluggish leading to several side products along with the desired polymer in a low yield. Another attempt towards the microwave assisted synthesis of polymer **9** was made when polymer bound catalyst $Pd(OAc)_2$-PEG was used. Unfortunately that was attempt was also not successful, leading to the formation of very little 9 in the reaction. Finally reaction of monomers **6** and **9** in water in a microwave reactor was tried where [1,1'-bis-(diphenylphosphino)-ferrocene]palladium(II) dichloride was used as a catalyst and tetra-n-butylammonium bromide as a phase transfer catalyst. [(J. Org. Chem, Vol. 68, No. 14, 2003 5667)]. This attempt was reasonably successful as the desired polymer **9** could be isolated in 52% yield (Scheme III).

Having not satisfactory synthesis from the microwave assisted approach, we next tried the synthesis of polymer **9** by using conventional method. Therefore, monomers **6** and **9** were treated under Suzuki-Miyuria conditions, using, 1'-bis(diphenylphosphino)ferrocene]-palladium(II) dichloride as a catalyst and K_2CO_3 as a base. The reaction yielded the desired polymer 9 in very high yield (89%) (Scheme III).

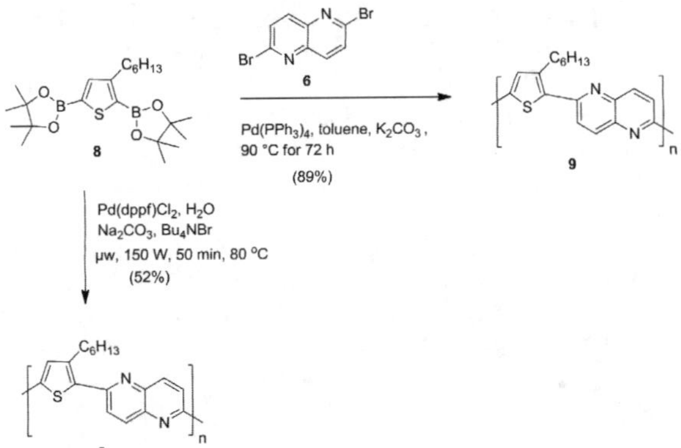

Scheme III. synthesis of Poly[1,5-naphthyridine-(3-hexylthophene)] (**9**)

The characteristic FTIR and proton NMR spectra of Poly[1,5-naphthyridine-(3-hexylthiophene)] (9) are given below in Figures 10-11.

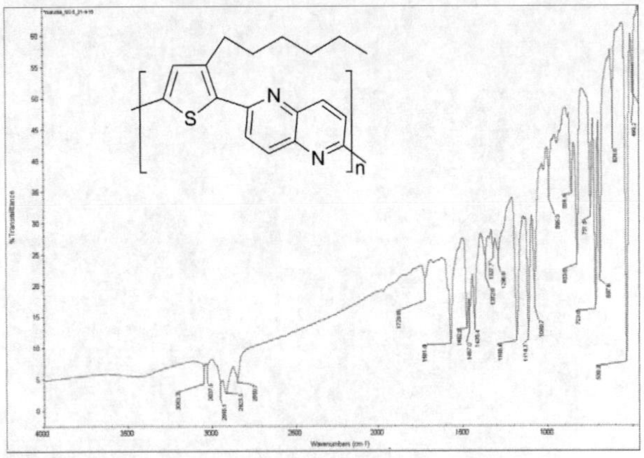

Figure 10. IR spectra of Poly[1,5-naphthyridine-(3-hexylthiophene)] (9).

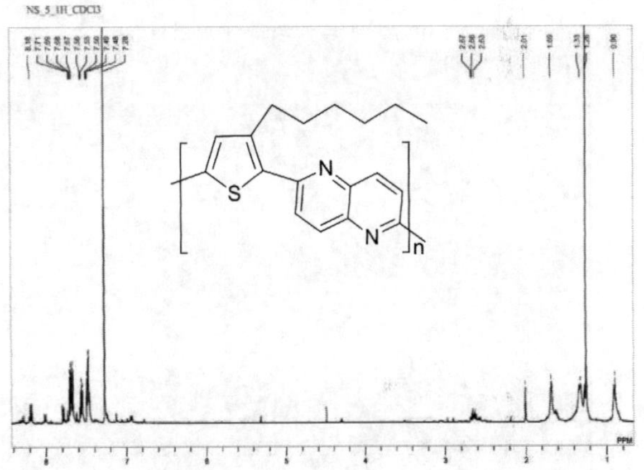

Figure 11. Proton NMR spectra of Poly[1,5-naphthyridine-(3-hexylthiophene)] (9).

Field emission scanning electron microscope (FESEM) was used to characterize the morphology and microstructure of the synthesized polymer. The FESEM confirms the uniform structure of the synthesized polymer, Poly[1,5-naphthyridine-(3-hexylthiophene)] (9), as shown in figures 12 a-c.

Raman spectrum of Poly[1,5-naphthyridine-(3-hexylthiophene)] is shown in Figure 13. The peaks corresponding to wave number 1465, 1369, and 1099 cm-1 can be seen which corresponds to thiophene unit. The C–C bonding stretching peak at ~ 1272. C-N stretching causes some shifts in peak position. The band at 913 cm-1 is due to the ring-breathing mode of the N=C bond. The bands at 1099 cm-1 correspond to ring mode of C=N bond. The positions of these peaks are in good agreement with the adsorption of S-C, C=C and N=C surfaces reported in previous SERS work.

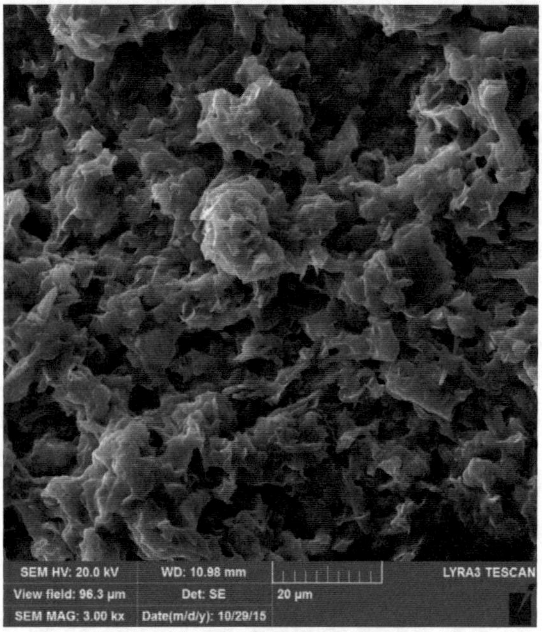

Figure 12-a. FESEM of Poly[1,5-naphthyridine-(3-hexylthiophene)] (9).

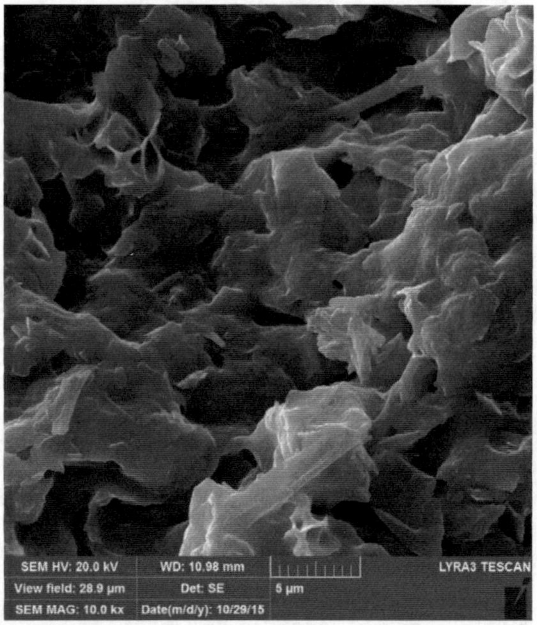

Figure 12-b. FESEM of Poly[1,5-naphthyridine-(3-hexylthiophene)] (9).

Figure 12-c. FESEM of Poly[1,5-naphthyridine-(3-hexylthiophene)] (9).

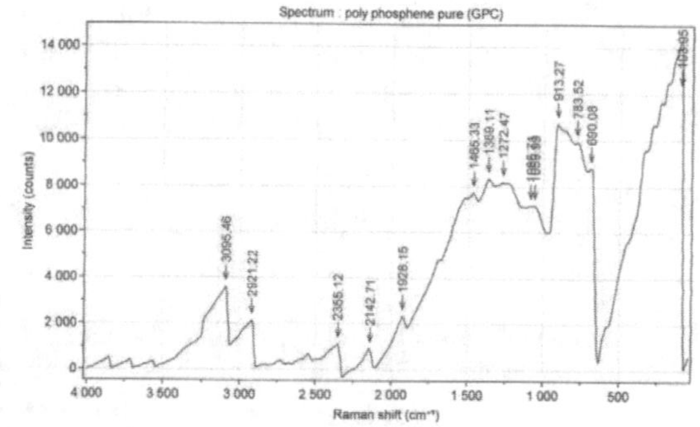

Figure 13. Raman spectra of Poly[1,5-naphthyridine-(3-hexylthiophene)] (9).

The cyclic voltammetry studies of polymer **9** are summarized in figures 14 and 15.

Figure 14 indicates CV results obtained for **9** as a bulk form in H_2SO_4 solution. The concertation of the polymer **9** in the solution was not well defined because the polymer had a low solubility in aqueous solutions. A well-defined oxidation peak at around 0.01 V is due to the slow transport of counter anions [1, 2]. The irreversible peak can be attributed to the conjugated system that has been disturbed by the lone pairs of the two nitrogen atoms. On the other hand, the sulfur acts as an electron donor which makes the system more electron-rich. This can also be explained due to the absence of the reduction peak.

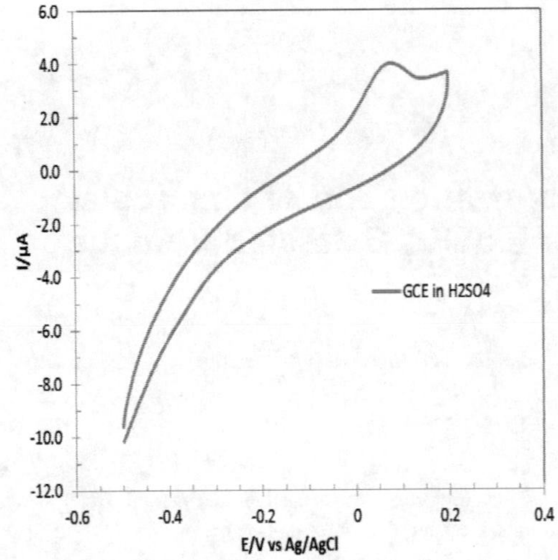

Figure 14. Cyclic voltammogram of PN_3HTh in 1.0 M H_2SO_4

Figure 15 shows the CV results obtained for polymer **9** as a thin film form in H_2SO_4 solution. The polymer was dissolved in acetonitrile (ACN) and boron trifluoride diethyl etherate ($BF_3.OEt_2$) before casting on a glassy carbon electrode. Two single oxidation peaks were obtained at around 0 V and 0.3 V for both cases. The potential scan started at -0.5 V and cycled to +0.5 V shows two-electron transport process which indicated the ability of p-doping. Different solvents can tune the transport properties of the polymer as can be seen from the two CVs where BF_3 exhibited enhanced transport properties over ACN.

It also showed electroactivity that can be improved further by addition/optimization of the functional groups on the skeleton of the polythiophenes structure. In order to make conjugated system at the same plane, the polythiophenes system should have structures where pi-electron can delocalized easily.

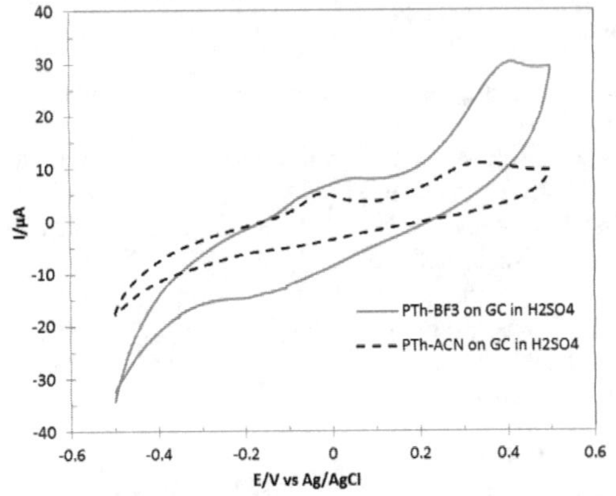

Figure 15. Cyclic voltammogram of PN_3HTh casted on GCE from BF_3 solution (solid line) and ACN solution (dash line) in 1.0 M H_2SO_4

Acknowledgement

The financial support provided by King Abdulaziz City for Science and Technology (KACST), Riyadh, Saudi Arabia through Project No. AR-34-36 and facility support provided by the King Fahd University of Petroleum & Minerals, Dhahran, Saudi Arabia for this work are gratefully acknowledged.

References

1. C. K. Chaing, C.R. Jr. Fincher, Y.W. Park, A. J. Heeger, H. Louis, E. J. Shirakawa, S. C. Gau, A. G. MacDiarmid, Physical Review Letters **39**, 1098 (1977)
2. J. Roncali, Chemical Reviews, **92**, 711 (1992).
3. A. D. Schluter, In Handbook of Conducting Polymers; 2 ed.; Skotheim, T. A. E., R.L.; Reynolds, J. R., Ed.; Marcel Dekker, INC. : New York, 209 (1998)
4. D. Curran, J. Grimshaw, S. Perera, Chemical Society Reviews, **20**, 391 (1991)
5. S. C. Moratti, In Handbook of Conducting Polymers; 2 ed.; Skotheim, T. A. E., R.L.; Reynolds, J. R., Ed.; Marcel, Dekker, INC. : New York, 343 (1998)
6. I. Osaka, E. G. Sauv, R. Zhang, T. Kowalewski, R. D. McCullough, Advanced Materials, **19**, 4160 (2007)
7. J. Liu, T. Tanaka, K. Sivula, A. P. Alivisatos, J. M. J. Frechet, Journal of the American Chemical Society, **126**, 6550 (2004)
8. Y. J. Cheng, S. H. Yang, C. S. Hsu, Chemical Reviews, **118**, 234 (2009)
9. M. Braunger, A. Barros, M. Ferreira, C. Olivati, Electrochimica Acta, **165**, 6 (2015)
10. P. Sivaraman, S. Mishra, A. Bhattacharry, A. Thakur, K. Shashidhara, A. Samui, Electrochimica Acta, **69**, 138 (2012)
11. M. H. Norman, N. Chen, Z. Chen, C. Fotsch, C. Hale, N. Han, R. Hurt, T. Jenkins, J. Kincaid, L. Liu, Y. Lu, O. Moreno, V. J. Santora, J. D. Sonnenberg, W. Karbon. Journal of Medicinal Chemistry, **43**, 22, 4288-4312 (2000)

MATEC Web of Conferences **49**, 03001 (2016)
DOI: 10.1051/matecconf/20164903001

Study of Fastness, UV Protection, Deodorization and Antimicrobial Properties of Silk Fabrics Dyed with the Liquids Extracted from the Gallnuts, Areca Nuts, and Pomegranate Peels

Jin Soun. Jung [a]

Department of Fashion Design, Semyung University, South Korea

Abstract. The purpose of this research is to study the fastness, UV-protection, deodorization, and antimicrobial properties of silk fabrics dyed with liquids extracted from the gallnuts, areca nuts, and pomegranate peels.

1 Introduction

In textile industry, to synthetic dyestuffs and pigments are widely used because of their various range of colours, better colour fastness properties and low prices [1]. However, synthetic dyestuffs and pigments are ruled out by many producers because of their toxicity and carcinogenic effect, being not bio-degradable as well ecological [2]. Recently, the textile finishing industry tends to restrict the use of such synthetic dyestuffs and pigments in order for human health and environmental purposes. As a result, the use of natural dye has begun to increase for their better properties as being bio-degradable, non-toxic, origination no problem to human health and waste water contaminant [3-5]. Natural dyes are environmental friendly, low toxic and less allergenic. Due to these advantages, over the last decade the use of natural dyes has gained momentum in food, pharmaceutical, cosmetic and textile dyeing industry [6]. For many years, scientists have investigated the deodorizing/aroma [7], insect-repellent [8], flame retardant [9], protection against to UV rays [10] of plants dyeing and usability in the textile industry. Unlike the synthetic dyes, colorants derived from the nature are thought to be safe because of their non-toxic, non-carcinogenic and biodegradable nature [11]. Natural dyes mainly consist of phenolic compounds which play an important role in plant growth and reproducibility. Many of them have antioxidant activity and are also considered as antibacterial and anti-inflammatory compounds. They have been widely used as herbal medicines as well as natural dyeing agents. Phenolic compounds based on their different chemical structure, are divided to groups corresponding to flavonoids, quinones, curcuminoids and tannins [12]. Tannin is an astrigent vegetable product found in awide variety of plant such as bark, wood, fruit pods, leaves, roots and plant galls. Tannins are defined as naturally occurring water soluble polyphenolic compounds of high molecular weight (about 500-3000) containing phenolic hydroxyl groups to enable them to form effective crosslinks between proteins and other macromolecules [13].

The purpose of this research is to study the fastness, UV-protection, deodorization, and antimicrobial properties of silk fabrics dyed with the liquids extracted from the gallnuts, areca nuts, and pomegranate peels contained tannins. The light, dry cleaning, rub, and perspiration fastness of the dyed silk fabrics was evaluated. The UV protection factor of the dyed silks with SPF calculated in wavelength range of 290-400 nm range. The deodorization activity was made from concentration of residual ammonia gas in a container. The antimicrobial activity of the dyed silks was measured against *Staphylococcus aureus* and *Klebsiella pneumoniae*.

2 Theoretical background

2.1 Gallnut

Gallnuts are outgrowths of plant tissues produced when irritants are released by the larvae of gall insects such as those of the Cynipidae family, the gall wasps. This extract contains the highest naturally occurring levels of tannin (gallotannin, 50-75%), as well as smaller molecules such as gallic acid and ellagic acid. Additionally, this extract is known to possess pharmaceutical properties, including anti-inflammatory, antibacterial, and detoxifying properties [14]. Figure 1 showed the chemical structure of tannin(Gallnut tannin) contained in gallnut and image of gallnut dried.

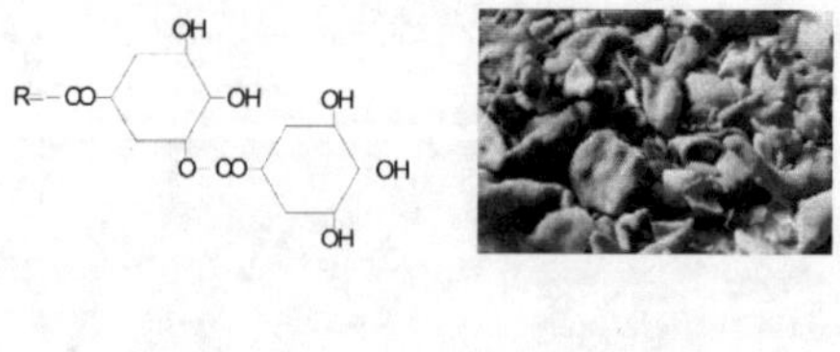

[a] Corresponding author: bobejin@semyung.ac.kr

Figure 1. Chemical structure of tannin(Gallnut tannin) contained in gallnuts and image of gallnuts dried.

2.2 Areca nut

Areca nut (*Areca catechu* L.), belonging to the family Palmae(or Arecaceae), native to Malaysia, widely cultivated in Indonesia, Sri Lanka, Hainan province, Guangdong province, Yunnan province and other places in Southeast Asia, is one of th most widely used South-China medicine resources [15]. Areca nut is popular chewable items used in traditional herbal medicine [16-18]. Areca nut exhibits multiple therapeutic properties like, aphrodisiac [19], antihypertensive [20, 21], wound healing [22], hypoglycemic [23, 24] and antidepressant [25]. It is one of the most commonly used drugs in the world, containing alkaloids, tannins, polyphenols, sugars, and lipids that have anthelmintic, antifungal, antibacterial, anti-inflammatory, and antioxidant activities [26]. Tannins are another characteristic component of Areca nut, and the main types are condensed tannins (also called proanthocyanidins). The main classes of tannins in A. catechu are the catechuins (Figure 2) [27-30].

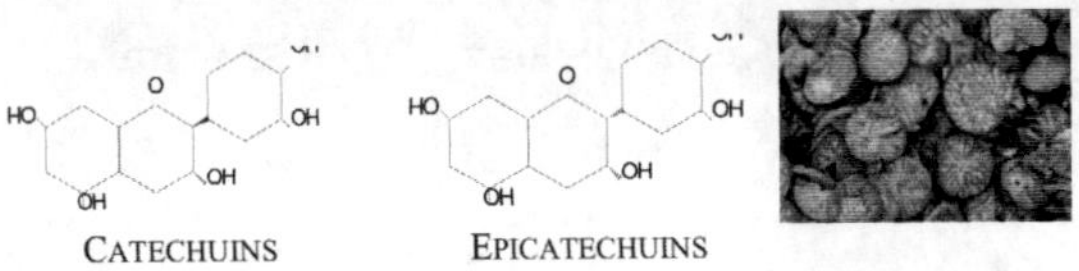

Figure 2. Chemical sructure of tannin(Catechuins and Epicatechuins) contained in areca nuts and image of areca nuts dried.

2.3 Pomegranate

Pomegranate (*Punica granatum* L.) belongs to the Punicacea family [31-33]. The cultivation of pomegranate is native to the Middle East and was later known in the Mediterranean. Pomegranate peels are rich in tannins [34-36]. They have been used traditionally for their medicinal properties as anticancer, anti-inflammatory, antioxidant and antithelminthic [37, 38] and for other purposes such as tanning, dyeing [39, 40] and heavy metal removal [41]. Pomegranate peels are characterized by an interior network of membranes comprising almost 26-30% of total fruit weight and are characterized by substantial amounts of phenolic compounds, including flavonoids (anthocyanins, catechins and other complex flavonoids) and hydrolysable tannins (punicalin, pedunculagin, punicalagin, gallic and ellagic acid) [42-44]. Gallic acid, ellagic acid and punicalagin, in addition to their free radical-scavenging properties, also possess antibacterial activites against intestinal flora, particularly enteric pathogens, i.e., Escherichia coli, Salmonella spp. Shigella spp., as well as Vibrio cholera [45-48]. Figure 3 showed the chemical structure of pomegranate tannin(Ellagic tannin) contained in pomegranate peels and image of pomegranate peels dried.

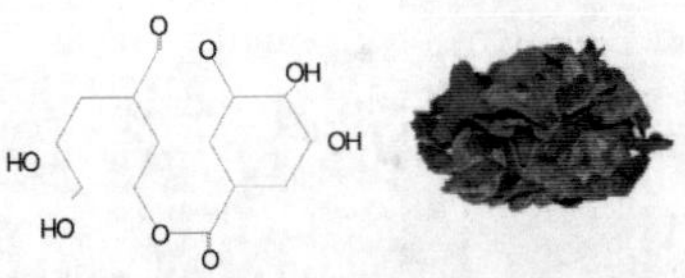

Figure 3. Chemical structure of pomegranate tannin(Ellagic tannin) contained in pomegranate peels and image of pomegranate peels dried.

3 Experimental materials and methods

3.1 Experimental materials

Silk: Silk used in this study was purchased from Testfabrics Inc. (West Pittston, PA)., and the characteristics are as shown on the Table 1. The silk was purchased from Testfabrics Inc. (West Pittston, PA).

Table 1. The Characteristics of silk

Fiber	Weave	Yarn Number		Fabric Counts (Threads/in.)		Weight (g/m^2)
		Warp	Weft	Warp	Weft	
Silk	Plain	21D	21D/2	56	39	26

Gallnut, Areca Nut, Pomegranate: Gallnuts and areca nuts were acquired from online from Cheongmyeong herbs(http://www.good1075.com), and pomegranates were purchased from a local market in Korea.

3.2 Experimental methods

UV-Vis/NIR Spectra: 1g dried gallnuts, areca nuts and pomegranate peels was added to 100 mℓ ethanol respectively, and they were extracted at room temperature for 24 hours, and filtered. The filtered extracts respectively were used as samples for UV-Vis analysis. The measurement of the UV-absorption characteristics was conducted in the range of 190-800 nm by using an ultraviolet-Visible/Near Infrared spectrophotometer (Varian Cary 5000).

FT-IR Spectra: The dried and grinded powers of gallnuts, areca nuts and pomegranate peels were analyzed with Fourier Transform Infrared Spectrometer (Bruker TENSOR27). Each samples were scanned registering the spectrum with 32 scans with a resolution 0f 4 cm^{-1} in the wave number range between 4000 and 600 cm^{-1}.

The extraction treatment of gallnut, areca nut and pomegranate peel: Gallnuts, areca nuts and pomegranate peels were extracted in liquor ratio of 1:20 at the boiling temperature for 20minutes. Each solutions were filtered with filter paper. The process was repeated 2 times. The liquid extraction combined first and second extract liquid was used as solution for dyeing.

Mordanting: Silk fabrics were mordanted by post-mordanting method using ferric mordant (0.2%), and liquor ratio for mordanting was kept at 1:30. Before the application of mordants, silk fabrics were soaked in distilled water. Water soaked silk fabrics were immersed in mordants solutions, and mordanted at 40 ℃ for 30minutes with constant stirring. Mordanted silk fabrics were rinsed with distilled water to remove superfluous mordants.

Dyeing: Before the application of dyeing, silk fabrics were soaked in distilled water. Firstly silk fabrics were dyed in liquor ratio of 1:30 at the boiling temperature for 20 minutes with constant stirring. Secondly, the samples were left in the fluid for one night. Thirdly, the samples were washed with 1500 ㎖ of cold distilled water (Three repetitions) then squeezed and dried at room temperature.

Color Fastness Tests: The light fastness of silk fabrics dyed was conducted on Fade-O-Meter(25-FR, Atlas Electric Devices Co. U.S.A) having water cooled Xenon Arc, as per test method KS K ISO B02:2005. The dry cleaning fastness of silk fabrics dyed was measured in Launder-O-Meter(Model LP2, Co. Atlas) as per the KS K ISO 105D01:2010, specification. The dry and wet rub fastness of silk fabric dyed s was tested using Rubbing-Crock Meter(CM-5, Atlas Electric Devices Co. U.S.A) as per the KS K ISO 0650:2011. The perspiration fastness of silk fabrics dyed was measured AATCC Perspiration Tester (PR-1, Atlas Electric Devices Co. U.S.A) as per test method KS K ISO 105E04:2010.

UV protection factor: UV-protection factor was tested using UV-Vis spectrophotometer(Varian Cary 5000) as per the KS K 0850:2014, o/d. Transmission measurements were made in 290-400 nm range with a 1 nm step. SPF was calculated according to:

$$SPF = \frac{\sum_{290}^{400} E_\lambda S_\lambda \Delta_\lambda}{\sum_{290}^{400} E_\lambda S_\lambda T_\lambda \Delta_\lambda}$$

where S_λ is the solar spectral irradiance at noon for a typical summer's day in central Italy, E_λ is the CIE erythermal spectral effectiveness, T_λ is the spectral transmittance of each fabric sample and T_λ is the wavelength step.

Deodorization activity: Deodorizationrate was calculated according to:

$$\text{Deodorizationrate}(\%) = \frac{Cb - Cs}{Cb} \times 100$$

where C_b is residual gas concentration of control after 2hours, Cs is residual gas concentration of specimen after 2hours.

Antimicrobial Activity: The antimicrobial ability of the dyed samples to impede microbial growth and retention was tested using *Staphylococcus aureus* and *Klebsiella pneumoniae* cultures, according to an established protocol to test the antibacterial of textiles (KS K 0693). Antimicrobial activity was calculated according to:

$$\text{Reduction bacteria}(\%) = \frac{B - A}{A} \times 100$$

where B is the number of bacteria recovered from the inoculated control specimen incubated for 18hours, A is the number of bacteria recovered from the inoculated treated test specimen incubated for 18hours.

4 Results

4.1 Spectroscopic analysis by UV - Vis/NIR spectra

Figure 4 shows the UV-Vis/NIR spectra of the ethanolic extraction solution of gallnuts, areca nuts and pomegranate peels in the range of 190-800 nm. As shown by Figure 4 and Table 2, two absorption bands are easily seen in the ranges from 190 to 250 nm, and from 250 to 300 nm, and another broad absorption band appears around 300-400 nm. Gallnuts presented two characteristic absorption maximum, λmax_1 around 217 nm and λmax_2 at 279 nm. Spectra of areca nuts classified as condensed tannin, presented two characteristic absorption maximum, λmax1 around 224 nm and λmax2 at 280 nm. Pomegranate peels absorbed with two λmax at 250 and 368 nm.

4.2 Spectroscopic analysis by FT-IR spectra

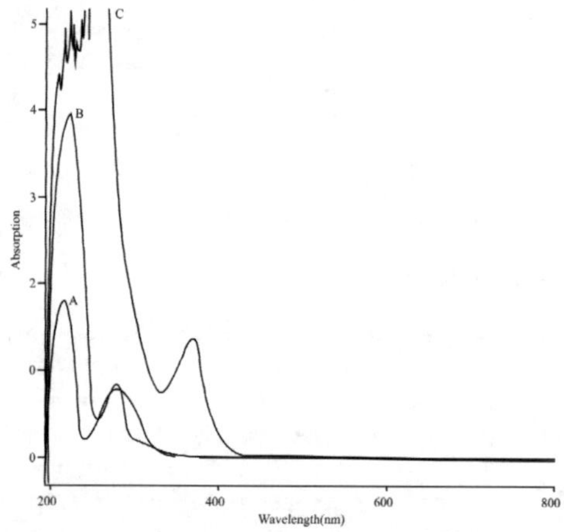

Figure 4. UV-VIS Spectra of ethanol extraction solution of gallnuts(A), areca nuts(B) and pomegranate peels(C)

Table 2. Wavelength and absorption of gallnuts, areca nuts and pomegranate peels

Samples	Wavelength(nm)	Absorption
Gallnut	217	1.813
	279	0.771
Areca nut	224	4.001
	280	0.856
Pomegranate peel	258	6.654
	368	1.366

4.3 Fastness properties

Fastness properties of silk fabrics dyed were given in Table 3. The samples showed mostly good light and dry cleaning fastness with 4 grade. Wet rub fastness was found to be relatively better than dry rub fastness. Perspiration fastness was all excellent grades 4~5 except for the 3~4 grades from discoloration by acidity and alkalinity.

Table 3. The fastness properties of silk fabrics dyed

Dyeing Fastness			Grade
Light Fastness			4
Dry Cleaning Fastness	Discoloration		4
	Solvent Contamination		4
Rub Fastness	Dry		2
	Wet		2~3
Perspiration Fastness	Acidity	Discoloration	3~4
		Contamination(Silk)	4
		Contamination(Cotton)	4~5
	Alkalinity	Discoloration	3~4
		Contamination(Silk)	4
		Contamination(Cotton)	4~5

4.4 UV Protection Rate

UV protection rate of dyed silk fabrics was shown in Table 4. UV-A protection rate of the samples in wavelength range of 290-400 nm showed 98.3%, and UV-B protection rate of the samples in wavelength range of 290 ~ 315 nm showed 98.4%. As described above, the samples appeared very good UV protection rate.

Table 4. UV protection rate of silk fabrics dyed

	UV Protection Rate (%)	
	UV-A	UV-B
Untreated Silk	-	-
Silk Fabric Dyed	98.3	98.4

4.5 Deodorization activity of silk fabric dyed

Table 5 showed deodorization activity of dyed silk fabrics. As outlined in Table 5, the samples appeared excellent deodorization activity over 99% even after 120min. test.

Table 5. Deodorization activity of silk fabric dyed

		Deodorization Activity (%)
Untreated Silk		-
Silk Fabric Dyed	30min.	over 99%
	60min.	over 99%
	90min.	over 99%
	120min.	over 99%

4.6 Antimicrobial ativity of silk fabric dyed

The antimicrobial activity of dyed silk fabrics against *Staphylococcus aureus* and *Klebsiella pneumoniae* was assessed. Table 6 showed the antiimicrobial activity of dyed silk fabrics. The samples appeared high antiimicrobial activity activity of 99.9% against *Staphylococcus aureus* and *Klebsiella pneumoniae*.

Table 6. Antimicrobial activity of silk fabrics dyed

	Reduction of Bacteria (%)	
	Staphylococcus aureus	*Klebsiella pneumoniae*
Untreated Silk	-	-
Silk Fabric Dyed	99.9	99.9

5 Conclusion

Among dyeing fastness of dyed silk fabrics, light and dry cleaning fastness showed 4 grade. Rub fastness was 2~3 grade. Perspiration fastness was 3~5 grade. The dyed silk fabrics in wavelength range of 290-400 nm appeared UV protection rate of 98.3%, and UV-B protection rate in wavelength range of 290 ~ 315 nm showed 98.4%. Deodorization activity of the dyed silk fabrics appeared over 99%. The dyed silk fabrics showed high antibacterial activity of 99.9% against *Staphylococcus aureus* and *Klebsiella pneumoniae*.

Acknowledgment

This work was supported by the Ministry of Education of the Republic of Korea and the National Research Foundation of Korea (NRF-2015S1A5A2A01009314).

References

1. L. G. Angelini, A. Bertoli, S. Rolandelli, L. Pistelli, Agronomic potential of Reseda Luteola L. as new crop for natural dyes in textiles production, Industrial Crops and Products, **17**, 199-207 (2003)
2. N. A. Ibrahim, A. R. El-Gamal, M. Gouda, F. Mahrous, A new approach for natural dyeing and functional finishing of cotton cellulose, Carbohydrate Polymers, **82**, 1205-1211 (2010)
3. T. Bechtold, M. A. Amalid, R. Mussak, Natural dyes for textile dyeing: A comparison of methods to assess the quality of Canadian golden rod plant material, Dyes and Pigments, **75**, 287-293 (2007)
4. A. Barka, A. Assabbanea, A. Nounah, L. Laanab, Y. L. Ichou, Removal of textile dyes from aqueous solutions by natural phosphate as a new adsorbent, Desalination, **235**, 264- 275 (2009)
5. S. Baliarsingh, A. K. Panda, J. Jena, T. Das, Exploring sustainable technique on natural dye extraction from native plants for textile: Identification of colourants, colourimetric analysis of dyed yarns and their antimicrobial evaluation, Journal of Cleaner Production, **37**, 257-264 (2012)

6. A. K. Samanta, P. Agarwal, Application of natural dyes on textiles, Indian Journal of Fibre Textile Research, **34**, 384-399 (2009)
7. W. Sricharussin, C. Sopajaree, T. Maneerung, N. Sangsuriya, Modification of cotton fabrics with b-cyclodextrin derivative for aroma finishing, Journal of Textile Insitute, **100**, 682-687 (2009)
8. M. M. M. Specos, J. J. Garcia, J. Tornesello, P. Marinoa, M. D. Vecchia, M. V. D. Tesoriero, L. G. Hermida, Microencapsulated citronella oil for mosquito repellent finishing of cotton textiles, Transactions of the Royal Society of Tropical Medicine Hygiene, **104**, 653-658 (2010)
9. L. Huang, M. Gerber, J. Lu, A. E. Tonelli, Formation of a flame retardantcyclo dextrin inclusion compound and its application as a flame retardant for poly (ethylene terephthalate), Polymer Degration and Stability, **71**, 279- 284 (2001)
10. D. Grifoni, L. Bacci, G. Zipoli, L. Albanese, F. Sabatini, The role of natural dyes in the UV protection of fabrics made of vegetable fibres, Dyes Pigments, **91**, 279-285 (2011)
11. M. Yusuf, A. Ahmad, M. Shahid, M. I. Khan, S. A. Khan, N. Manzoor, F. Mohammad, Assessment of colorimetric, antibacterial and antifungal properties of woollen yarn dyed with the extract of the leaves of henna (Lawsonia inermis), Journal of Cleaner Production, **27**, 42-50 (2012)
12. F. Alihosseini, G. Sun, Recent progresses in antibacterial dyes, H and PC Tod., **4**, 17-21 (2008)
13. K. Ramakrishnan, S. R. Selve, R. Shubha, Tannin and its analytical techniques, Indian Chemical Engineering Section A, **48**, 2, 88- 93 (2006)
14. Park AY, Kim IY, Song WS., The effect of gallnut mordanting on gromwell dyed silk fabric, Journal of the Korean Society of Clothing and Textiles, **33**, 256 - 265 (2009)
15. J. L. Huang, High-performance liquid chromatographic determi- nation of the alkaloids in betel nut, Journal of Chromatography, **475**, 447- 450 (1989)
16. M. S. Amudhan, V. H. Begum, K. B. Hebbar, A review on phytochemical and pharmacological potential of Areca catechu L. seed, International Journal Pharmaceutical Science and Research, **3**, 4151-4157 (2012)
17. Anonymous, Dictionary of Chinese Materia Medica, Science and Technology Press of Shanghai, China: Shanghai, 2525-2528, (1977)
18. C. D. Heatubun, J. Dransfield, T. Flynn, S. S. Tjitrosoedirdjo, J. P. Mogea, W. Baker, A monograph of the betel nu tpalms(Areca: Arecaceae) of East Malesia, Botanical Journal of the Linnean Society, **168**, 147-173 (2012)
19. S. A. Norton, Betel: Consumption and consequences, Journal of the American Academy of Dermatology, **37**, 81- 88 (1998)
20. J. Inokuchi, H. Okabe, T. Yamauchi, A. Nagamatsu, G. Nonaka, I. Nishioka, Antihypertensive substance in seeds of Areca catechu L., Life Sciences, **38**, 1375 - 1382 (1986)
21. Y. W. Xie, H. X. Xuc, H. Dongc, R. R. Fiscus, P. H. Paul, Role of nitric oxide in the vasorelaxant and hypotensive effects of extracts and purified tannins from Geum japonicum, Journal of Ethnopharmacology, **109**, 128- 133 (2007)
22. S. Azeez, S. Amudhan, S. Adiga, N. Rao, N. Rao, L.A. Udupa, Wound healing profile of Areca catechu extracts on different wound models in wistar rats, Kuwait Medical Journal, **39**, 1) 48- 52 (2007)
23. J. K. Grover, S. Yadav, V. Vats, Medicinal plants of India with anti-diabetic potential, Journal of Ethnopharmacology, **81**, 81- 100 (2002)
24. P. K. Mukherjee, K. Maiti, K. Mukherjee, P. J. Houghton, Leads from Indian medicinal plants with hypoglycemic potentials, Journal of Ethnopharmacology, **106**, 1- 28 (2006)
25. A. Dar, S. Khatoon, Behavioral and biochemical studies of dichloromethane fraction from the Areca catechu nut, Pharmacology Biochemistry and Behavior, **65**, 1, 1- 6 (2000)
26. G. W. Staples, R. F. Bevacqua, Areca catechu (betel nut palm), Available at http://www.webalice.it/siamseeds/Database/Areca-catechu-betel-nut (2006)
27. Y. T. Ma, F. L. Hsu, S. J. Lan, C. F. Chen, Tannins from betel nuts, Journal of Chinese Chemical Society, **4**, 77- 81 (1996)
28. G. I. Nonaka, F. L. Hsu, I. Nishioka, Structures of dimeric, trimeric, and tetrameric procyanidins from Areca catechu L., Journal of Chemical Society Chemistry Communication, **9**, 781- 783 (1981)
29. Anonymous, Chinese Material Medica, Science and Technology Press of Shanghai, China: Shanghai, **8**, 439- 648 (1999)
30. W. Q. Yang, H. C. Wang, W. J. Wang, Y. Wang, X. Q. Zhang, W. C. Ye, Chemical constituents fromthefruitsof Areca catechu, Journal of Chinese Medical Materials, **35**, 400- 402 (2012)
31. V. Akbarpour, K. Hemmati, M. Sharifani, Physical and chemical properties of pomegranate (Punica granatum L.) fruit in maturation stage, American-Eurasian Journal of Agricultural & Environmental Sciences, **6**, 411- 416 (2009)
32. M. Ozgen, C. Durgac, S. Serc, C. Kaya, Chemical and antioxidant properties of pomegranate cultivars grown in the Mediterranean region of Turkey, Food Chemistry, **111**, 7703- 7706 (2008)
33. R. A. Newman, E. P. Lansky, M. L. Block, Pomegranate: The Most Medicinal Fruit, 1st ed. W. Roberta, Ed. USA: Waddell (2007)
34. N. S. Al-Zoreky, Antimicrobial activity of pomegranate (Punica granatum L.) fruit peels, International Journal of Food Microbiology, **134**, 244- 248 (2009)
35. T. B. Machado, A. V. Pinto, M. C. F. R. Pinto, I. C. R. Leal, M. G. Silva, A. C. F. Amaral, R. M. Kuster, K. R. Nett-dosSantos, In vitro activity of Brazilian medicinal plants, naturally occurring naphthoquinones and their analogues against methicillin-resistant Staphylococcus aureus,

International Journal of Antimicrobial Agents, **21**, 279- 284 (2003)

36. S. Voravuthikunchai, A. Lortheeranuwat, W. Jeeju, T. Sririrak, S. Phongpaichit, T. Supawita, Effective medicinal plants against enterohaemorrhagic Escherichia coli O157:H7, Journal of Ethnopharmacology, **94**, 49- 54 (2004)
37. E. A. Hayouni, K. Miled, S. Boubaker, Z. Bellasfar, M. Abedrabbad, H. Iwaskie, H. Okue, T. Matsuie, F. Limama, M. Hamdi, Hydroalcoholic extract based-ointment from Punica granatum L. peels with enhanced in vivo healing potential on dermal wounds, Phytomedicine, **18**, 976- 984 (2011)
38. G. Mousavinejad, Z. Emam-Djomeh, K. Rezaei, M. H. H. Khodaparast, Identification and quantification of phenolic compounds and their effects on antioxidant activity in pomegranate juices of eight Iranian cultivars, Food Chemistry, **115**, 1274- 1278 (2009)
39. A. Adeel, I. A. Ali Shaukat Bhatti, F. Zsila, Dyeing of cotton fabric using pomegranate (Punica granatum) aqueous extract, Asian Journal of Chemistry, **21**, 3493- 3499 (2009)
40. J. U. Lloyd, Punica granatum, second ed. Chicago (consulted in: http://swsbm.henriettesherbal.com/ManualsOther/Punica granatum Lloyd (April 2001)
41. T. S. Najim, S. A. Yassin, Removal of Cr (VI) from aqueous solution using modified pomegranate peel: equilibrium and kinetic studies, European Journal of Chemistry, **6**, S129- S142 (2009)
42. F. Afaq, M. Saleem, C. G. Krueger, J. D. Reed, H. Mukhtar, Anthocyanin- and hydrolyzable tannin-rich pomegranate fruit extract modulates MAPK and NF-kappaB pathways and inhibits skin tumorigenesis in CD-1 mice, International Journal of Cancer, **113**, 423-433 (2005)
43. P. S. Negi, G. K. Jayaprakasha, B. S. Jena, Antioxidant and antimutagenic activities of pomegranate peel extracts, Food Chemistry, **80**, 393-398 (2003)
44. M. Zahin, F. Aqil, I. Ahmad, Broad spectrum antimutagenic activity of antioxidant active fraction of Punica granatum L. peele xtracts, Mutation Research, **703**, 99-107 (2010)
45. M. Aviram, N. Volkova, R. Coleman, M. Dreher, M. K. Reddy, D. Ferreira, M. Rosenblat, Pomegranate phenolics from the peels, arils, and flowers are 9antiatherogenic:studies in vivo in atheroscleroticapo lipoproteine-deficient (E0) mice and in vitro in cultured macro phages and lipoproteins, Journal of Agricultural and Food Chemistry, **56**, 1148- 1157 (2008)
46. J. Lu, Y. Wei, Q. Yuan, Preparative separation of punicalagin from pomegranate husk by high-speed countercurrent chromatography, Journal of Chromatography B, Analytical Technologies in the Biomedical and Life Sciences, **857**, 175-179 (2007)
47. V. Pai, T. R. Chanu, R. Chakraborty, B. Raju, R. Lobo, M. Ballal, Evaluation of the antimicrobial activity of Punica granatum peel against the enteric pathogens: an in vitro study, Asian Journal of Plant Science and Research, **1**, 2, 57- 62 (2011)

MATEC Web of Conferences **49**, 04001 (2016)
DOI: 10.1051/matecconf/20164904001

Hydrodynamic and Mass Transfer Characteristics of Annulus-Rising Airlift Reactor – The Effect of Reactor Scale

Mei Han [1,2,a], Arto Laari [1], Tuomas Koiranen [1]

[1] *Lappeenranta University of Technology, Lappeenranta, Finland*
[2] *Tianjin University of Science and Technology, Tianjin, China*

Abstract. Hydrodynamic and mass transfer characteristics of an annulus-rising airlift reactor (AR_ALR) were investigated with experimental and CFD simulation methods. A three-phase Eulerian model was developed to simulate the AR_ALR in three flow regimes. 3D steady state CFD simulations were performed under different gas superficial velocities (*Ug*). Good agreements on gas holdup and volumetric mass transfer coefficient were obtained over the range of the studied *Ug*. The three flow regimes in AR_ALR were captured well and are similar to those observed in experiments. The developed CFD model can be used to predict mass transfer and gas holdup in AR_ALRs at different scales.

1 Introduction

Airlift loop reactors (ALRs) have received more and more attention in chemical, petrochemical and biochemical industries. They are used in fermentation, waste water purification, hydrogenation and exhaust-gas treatment [1, 2]. ALRs have some advantages over bubble columns such as enhanced mixing, mass and heat transfer, and suspension of particles with low energy consumption. The mild and constant shear environment in ALRs, contrary to mixing tanks, is preferable for bioprocesses with fragile particles [3].

ALRs usually are composed of a riser with a gas feed, and a downcomer where the liquid phase flows downwards. An enlarged expansion, called gas separator, is often placed on the top of the column in order to achieve good separation of gas and liquid phases. Generally, two types of ALRs, an internal-loop airlift reactor (IL-ALR) and an external-loop airlift reactor (EL-ALR), are classified according to the arrangement of the riser and the downcomer. For IL-ALR, two operating modes can be used which are the gas-feeding in the draft tube (center-rising airlift reactor, CR_ALR) and the gas feeding in the annulus (annulus-rising airlift reactor, AR_ALR) modes.

A large number of studies on CR_ALRs have been reported and great achievements have been obtained with experimental [4-8] or simulation [9-14] methods. However, studies on AR-ALRs are quite scarce [15-18] although Koide *et al.* [17, 18] have shown in their experimental studies that AR_ALRs have in fact higher gas holdups and mass transfer rates than CR_ALRs, especially for liquid phase systems which have frothing properties. The anti-frothing ability of the AR_ALRs is higher than CR_ALRs due to much more bubbles being entrained into the downcomer. Moreover, most of the studies, typically in bioprocesses, have been performed in bench-scale (1 – 10 L) ALRs [15] and are limited to the specified aeration rates [2]. The yield of desired products may be lower than expected even in pilot-scale applications [19] due to the effects of the reactor scale, aeration and agitation. It is therefore necessary to investigate the effect of aeration and scale on hydrodynamics and mass transfer in AR_ALR so that the performance of the reactor would satisfy the process needs as well as possible.

The aim of this work was to investigate the effects of aeration and reactor scale on hydrodynamics and mass transfer characteristics of AR_ALRs in order to promote their industrial application. This study on a laboratory AR_ALR was, therefore, conducted with experiments and CFD simulations for different gas superficial velocities. A three-phase Eulerian model, two gas phases with different bubble size and a liquid phase, was developed to predict the mass transfer performance of the AR_ALR. Furthermore, the effect of reactor scale was predicted and compared with experimental studies from the literature.

2 Experiments

The schematic diagram of the experimental setup used in the studies is shown in Figure 1.

[a] Corresponding author: mei.han@lut.fi

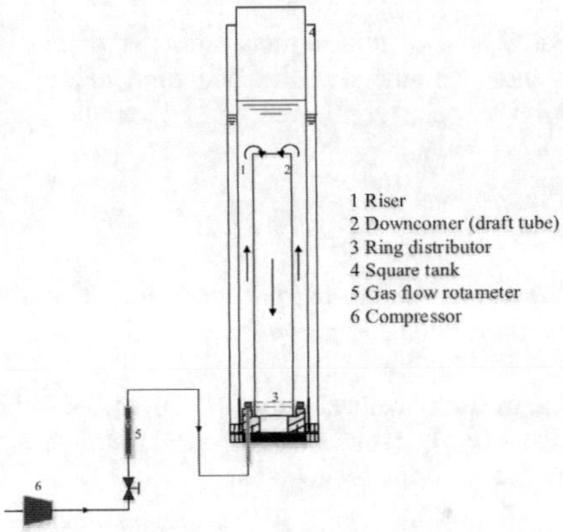

Figure 1. Schematic diagram of the experimental setup

The laboratory-scale AR_ALR consists of two cylinders made of acrylic resin with a thickness of 0.005 m. A plastic ring gas distributor with 36 pores of 0.5 mm diameter was installed at the annulus of the outer cylinder 0.07 m above the bottom. Table 1 gives the dimensions of the AR_ALR in details.

Table 1. The dimensions of the AR_ALR.

Dimension, m	Riser	Downcomer
diameter	0.15	0.10
height	1.20	0.60

Compressed air was used as the dispersed phase and it was operated in once-through mode. Tap water was the continuous phase and it was used as a batch. The unaerated liquid height was 0.8 m. All the experiments were carried out at room temperature and atmospheric pressure. The operating gas superficial velocity, U_g, was calculated based on the cross-sectional area of the riser and is listed in Table 2.

Table 2. The dimensions of the AR_ALR.

Gas superficial velocity	Value, cm/s				
U_g	0.41	0.65	1.18	1.74	2.55

The overall gas holdup, $<a_{g,overall}>$, in the column was measured by the height expansion method and calculated by Eq. (1).

$$\langle a_{g,overall} \rangle = (H_m - H_l)/H_m \qquad (1)$$

where H_m and H_l are the height of the mixture after aeration and the height of the liquid before aeration, respectively.

The volumetric mass transfer coefficient, k_La, was determined by using a dynamic oxygen absorption method. The dissolved oxygen concentration was monitored and recorded using a MARVET BASIC 2000 Dissolved Oxygen Meter (Tallinn, Estonia) with the oxygen sensor located at 0.6 m above the gas distributor. Nitrogen was first sparged into the AR-ALR until the dissolved oxygen concentration was close to zero level and steady-state hydrodynamic conditions were reached. Then, nitrogen flow was shut down and compressed air feed was started. A pressure valve and a digital mass-flow controller were used to keep the nitrogen and the compressed air feeds at constant level in order not to interrupt the flow fields. Sufficient time was given in each experimental run to strip out the dissolved nitrogen and to saturate the liquid phase with oxygen to the maximum level. To calculate the volumetric oxygen mass transfer coefficient, it is assumed that the liquid phase is perfectly mixed and that oxygen depletion from the gas bubbles is negligible. The rate of oxygen dissolution in the liquid phase can be described by Eq. (2),

$$dC_L/dt = k_L\alpha_L(C^* - C_L) \qquad (2)$$

where k_La_L is the volumetric mass transfer coefficient per unit volume of liquid in the AR_ALR. C_L is the dissolved-oxygen concentration in the liquid and C^* is the oxygen saturation concentration at the gas-liquid interface.

The sensor of the dissolved oxygen meter has a time constant, k_{sensor}, corresponding to the delay in recording due to the inherently finite response time. The oxygen concentration value indicated by the sensor, C_{sensor} satisfies [20],

$$dC_{sensor}/dt = k_{sensor}(C_L - C_{sensor}) \qquad (3)$$

where C_L is the concentration of dissolved oxygen. In the experiments to determine the value of k_{sensor}, two beakers with water saturated with nitrogen and oxygen were used, respectively. A value of 0.0625 1/s was obtained for the ksensor by making a regression of the C_{sensor}/C_L values versus time. Analytical solution of (2) along with the sensor dynamics given by Eq. (3), yields [21],

$$C_{sensor}/C^* = 1 - k_{sensor}e^{-k_La_Lt}/(k_{sensor} - k_La_L) + k_La_Le^{-k_{sensor}t}/(k_{sensor} - k_La_L) \qquad (4)$$

The volumetric mass transfer coefficient per volume of the reactor can then be determined by Eq. (5):

$$k_La = k_La_L(1 - \langle a_{g,overall} \rangle) \qquad (5)$$

3 Simulation

3.1 Governing equations

A three-phase Eulerian model was developed to simulate the hydrodynamics and mass transfer of AR-ALRs. This model includes two gas dispersed phases and one liquid continuous phase. Each gas phase has a uniform bubble size but a different bubble diameter value. The large bubble phase (g1) and the small bubble phase (g2) were defined according to measured bubble sizes. The diameter varies from 0.005 to 0.008 m for the large bubble class and from 0.001 to 0.003 m for the small bubble class corresponding to U_g. Interaction between g1 and g2 was not considered and no bubble coalescence or breakage was involved. The momentum transfer ($\mathbf{M}_k$) between the gas phases and the liquid phase was formulated by taking the drag ($\mathbf{F}_{D,k}$), lift ($\mathbf{F}_{L,k}$), wall lubrication ($\mathbf{F}_{W,k}$) and turbulent dispersion forces ($\mathbf{F}_{T,k}$) into account. The turbulence of the liquid phase was resolved by the dispersed standard k - ε turbulent model. The governing equations are listed in Table 3 in detail.

Table 3. The developed three-phase Eulerian model

Item	Model
Continuity	$\nabla\cdot(\alpha_k\rho_k\mathbf{u}_k)=0, k=l,g1,g2$
Momentum	$\nabla\cdot(\alpha_l\rho_l\mathbf{u}_l\mathbf{u}_l)=-\alpha_l\nabla P'+\nabla\cdot\left(\alpha_l\mu_{eff,l}\left(\nabla\mathbf{u}_l+\nabla\mathbf{u}_l^T\right)\right)+\rho_l\alpha_l\mathbf{g}+M_l$
	$\nabla\cdot(\alpha_{gi}\rho_{gi}\mathbf{u}_{gi}\mathbf{u}_{gi})=-\alpha_{gi}\nabla P'+\nabla\cdot\left(\alpha_{gi}\mu_{eff,gi}\left(\nabla\mathbf{u}_{gi}+\nabla\mathbf{u}_{gi}^T\right)\right)+\rho_{gi}\alpha_{gi}\mathbf{g}+M_{gi}$ $i=g1,g2$
	$M_l=-(M_{g1}+M_{g2})=\sum_{i=1}^{2}\mathbf{F}_{D,gi}+\sum_{i=1}^{2}\mathbf{F}_{L,gi}+\sum_{i=1}^{2}\mathbf{F}_{W,gi}+\sum_{i=1}^{2}\mathbf{F}_{T,gi}$
$\boldsymbol{F}_{D,gi}$	$\mathbf{F}_{D,gi}=3/4\,\alpha_{gi}\alpha_l\rho_l\,C_{D,gi}/d_{gi}\lvert\mathbf{u}_{gi}-\mathbf{u}_l\rvert(\mathbf{u}_{gi}-\mathbf{u}_l)$
	$C_{D,g1}=\max\left[\min\left(24/\mathrm{Re}_{r,g1}\left(1+0.15\mathrm{Re}_{r,g1}^{0.687}\right),72/\mathrm{Re}_{r,g1}\right),8/3\left(Eo_{g1}/\left(Eo_{g1}+4\right)\right)\right]$
	$C_{D,g2}=\begin{cases}24\left(1+0.15\mathrm{Re}_{r,g2}^{0.687}\right)/\mathrm{Re}_{r,g2}, & \mathrm{Re}_{r,g2}\le 1000\\ 0.44, & \mathrm{Re}_{r,g2}>1000\end{cases}$
$\boldsymbol{F}_{L,gi}$	$\mathbf{F}_{L,gi}=-C_{L,gi}\alpha_{gi}\rho_l(\mathbf{u}_{gi}-\mathbf{u}_l)\times(\nabla\times\mathbf{u}_l)$
	$C_{L,gi}=\begin{cases}\min(0.288\tanh n(\mathrm{Re}_{r,gi}),f(Eo_{gi})), & Eo_{gi}\le 4\\ f\left(Eo_{gi}\right), & 4< Eo_{gi}\le 10\\ -0.27, & 10< Eo_{gi}\end{cases}$
	$f(Eo_{gi})=0.00105Eo_{gi}^{'3}-0.0159Eo_{gi}^{'2}-0.0204Eo_{gi}^{'}+0.474$
$\boldsymbol{F}_{T,gi}$	$\mathbf{F}_{T,gi}=-\rho_l C_T\nabla\alpha_l,\ C_T=0.1$
$\boldsymbol{F}_{W,gi}$	$\mathbf{F}_{W,gi}=\alpha_{gi}\rho_l(\mathbf{u}_{slip,gi}-(\mathbf{u}_{slip,gi}n_w)n_w)/d_{gi}\times\max(C_{W1}+C_{W2}\,d_{gi}/y_W,0), C_{W1}=-0.01, C_{W2}=0.05$
k - ε	$\nabla\cdot(\rho_l\alpha_l k_l\mathbf{u}_l)=\nabla\cdot\left(\alpha_l\left(\mu_{lam,l}+(\mu_{tl}+\mu_{tb})/\sigma_k\right)\nabla k_l\right)+\alpha_l(G_{k,l}-\rho_l\varepsilon_l)+\alpha_l S_{k,l}$
	$\nabla\cdot(\rho_l\alpha_l\varepsilon_l\mathbf{u}_l)=\nabla\cdot\left(\alpha_l\left(\mu_{lam,l}+(\mu_{tl}+\mu_{tb})/\sigma_\varepsilon\right)\nabla\varepsilon_l\right)+\alpha_l\frac{\varepsilon_l}{k_l}(C_{\varepsilon1}G_{k,l}-C_{\varepsilon2}\rho_l\varepsilon_l)+\alpha_l S_{\varepsilon,l}$
	$\mu_{eff,l}=\mu_{lam,l}+\mu_{tl}$, $\mu_{eff,gi}=\rho_g/\rho_l\,\mu_{eff,l}$ C_μ = 0.09; σ_k =1.00; σ_ε =1.3; $C_{\varepsilon1}$ = 1.44, $C_{\varepsilon2}$ = 1.92

3.2 Mass transfer model

Mass transfer rate between gas and liquid phase is an important factor in multiphase reactor performance. Interfacial mass transfer is usually expressed by volumetric mass transfer coefficient k_La. Four different categories of mass transfer models have been reported in [13]. These are (1) phenomenological correlations and/or models; (2) spatial models; (3) time models and (4) combined film-penetration models. Huang *et al.* [13] compared several typical mass transfer models for one ALR and recommended the one time model proposed earlier by Cocke *et al.* [22] based on the penetration theory, as shown in Eq. (6).

$$k_La=12\alpha_g/d_b\sqrt{D_LU_{slip}/\pi d_b} \tag{6}$$

Equation (7) and (8), in the current study, were employed to calculate the k_La between the liquid and each bubble phase, respectively.

The overall k_La, the sum of $(k_La)_{g1}$ (large bubble size class) and $(k_La)_{g2}$ (small bubble size class), was validated against the experimental results.

$$(k_La)_{g1}=12\alpha_{g1}/d_{g1}\sqrt{D_LU_{slip,g1}/\pi d_{g1}} \tag{7}$$

$$(k_La)_{g2}=12\alpha_{g2}/d_{g2}\sqrt{D_LU_{slip,g2}/\pi d_{g2}} \tag{8}$$

3.3 Numerical details

To reduce computational expense, one quarter of the 3-dimensional AR_ALR was taken as the computational domain. The configuration of the AR_ALR was generated according to the experimental equipment. The velocity inlet boundary was applied to the upper face of the gas distributor, where the gas inlet velocity and its volume fraction were specified according to the experiments. On the top domain surface, pressure outlet boundary condition was set. The symmetry boundaries were used for the two symmetrical surfaces of the quarter of the AR_ALR. The boundary conditions were a non-slip condition for the liquid and gas phase on all the reactor walls.

3D steady state simulations were carried out with the commercial software ANSYS Fluent 14.5. For all the simulations, the phase-coupled SIMPLE algorithm was used for the solution of the pressure-velocity coupling. The second order upwind scheme was used for the momentum, turbulence kinetic energy and dissipation rate equations and the first-order scheme for the volume fraction equation. The residual convergence criterion for all the variables was set to 10^{-4}. The mass balance of the gas phase and the volume-weighted average gas holdup for each case were additionally used to examine the convergence for all simulations.

4 Result and discussion

4.1 Mesh independence

The influence of the mesh size on the numerical solutions was examined with one coarse mesh case and another finer mesh. The coarse mesh had 43,824 hexahedron elements with the maximum edge size of 5 mm. The number of the cells of the fined one was 400,896 with the maximum edge size of 2.5 mm. The simulations were performed for both cases at gas superficial velocity U_g 1.18 cm/s.

The relative discrepancy between the coarse and the finer mesh cases is 1.58% for the overall gas holdup and 0.81% for the $k_L a$. Figure 2 shows the radial profile of the axial liquid velocity for both cases, where the difference is around 10%. It can be seen that the coarse mesh is sufficient to obtain mesh independent solution. Therefore, it was used to conduct all the subsequent simulations.

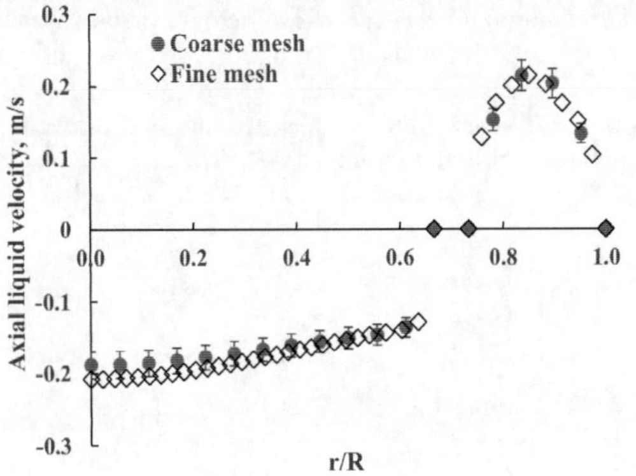

Figure 2. Influence of mesh size on radial profile of axial liquid velocity

4.2 Simulation and validation of the lab-scale AR_ALR

The simulated overall gas holdup agrees well with the experimental results over the range of the studied U_g, as shown in Figure 3. The overall gas holdup increases with the increase of U_g.

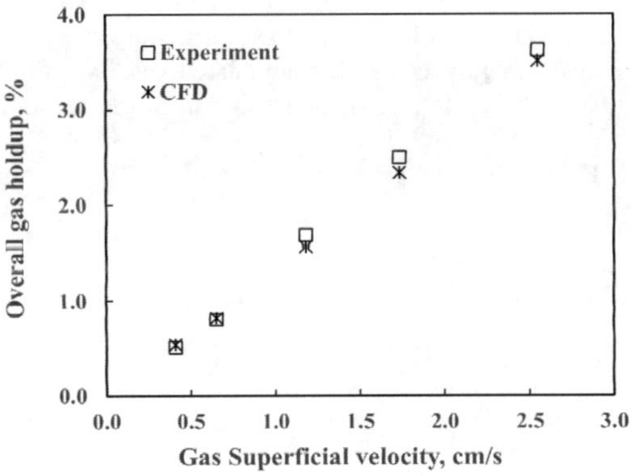

Figure 3. Overall gas holdup at different U_g : square-experiments; star-CFD.

Three flow regimes can be distinguished in ALRs depending how far the gas bubbles reach in the reactor. These are the "no gas entrainment" regime (regime I), "gas entrainment, but no recirculation" regime (regime II) and "complete gas recirculation" regime (regime III). As shown in Figure 4 (a), (b) and (c), all the three flow regimes, which were also observed in the experiments, were captured with the developed model at different gas superficial velocities.

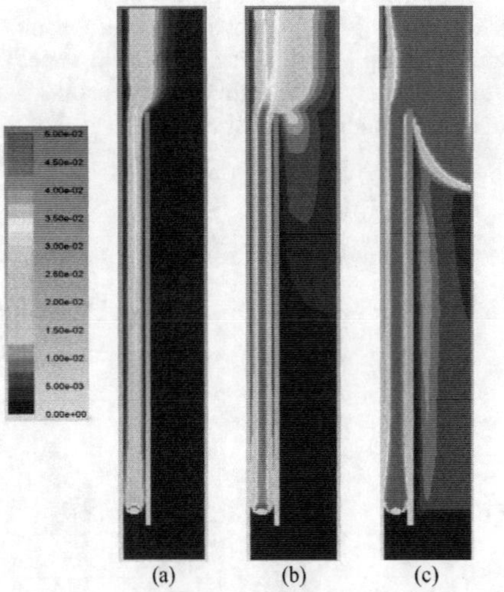

Figure 4. Gas holdup profiles for three regimes:

(a) U_g =0.65 cm/s (b) U_g =1.18 cm/s (c) U_g =2.55 cm/s.

The volumetric mass transfer coefficient was validated against the experimental results under different Ug, as shown in Figure 5. In overall, good agreement was obtained between the simulations and experiments although some discrepancy can be seen at Ug = 1.18 cm/s. This is probably caused by more complex hydrodynamic characteristics at the transition from homogeneous to heterogeneous flow.

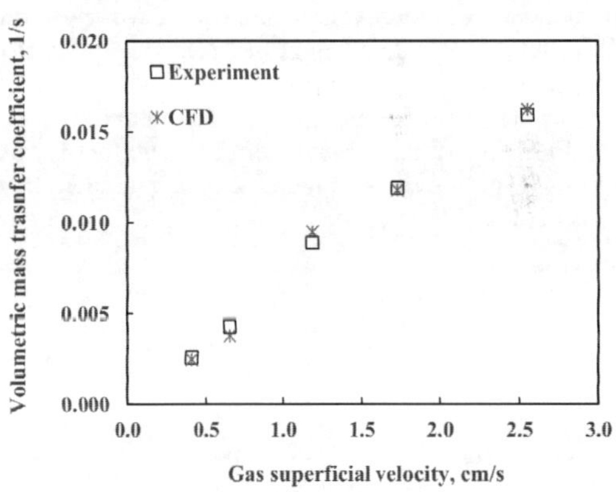

Figure 5. Mass transfer at different U_g : square-experiments; star-CFD.

The local volumetric mass transfer coefficient from simulations is shown in Figure 6 for three different flow regimes. The maximum volumetric mass transfer coefficient appears at the inlet of the downcomer due to substantial accumulation of bubbles. It can be seen that the radial profile of $k_L a$ is 'core-peaking' in the riser and 'wall-peaking' in the downcomer (except for the case (a) of 'no bubble entrained into the downcomer flow regime'), which is in accordance with the gas holdup profile also. Only small bubbles can be entrained further into the downcomer. These small bubbles are located close to the wall of the downcomer. These results suggest that the developed three-phase Eulerian model can be used to accurately predict the hydrodynamics and mass transfer of the studied AR_ALR.

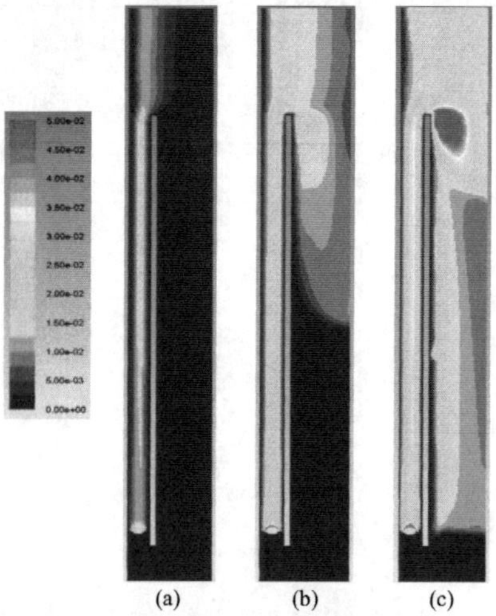

Figure 6. Mass transfer profiles for three regimes:

(a) U_g =0.65 cm/s (b) U_g =1.18 cm/s (c) U_g =2.55 cm/s.

4.3 The scale effect simulated by CFD

The effect of reactor scale on the hydrodynamics and mass transfer of AR_ALR was investigated with CFD simulations. 3-D steady-state simulation was performed for an AR_ALR with a height of 4.5 m. The ratio of the cross-sectional area of the riser to that of the downcomer was maintained the same as in the laboratory AR_ALR. The dimension of the computational domain for the larger AR_ALR is shown in Figure 7. The number of the elements was 773,652 with the maximum edge of 5 mm.

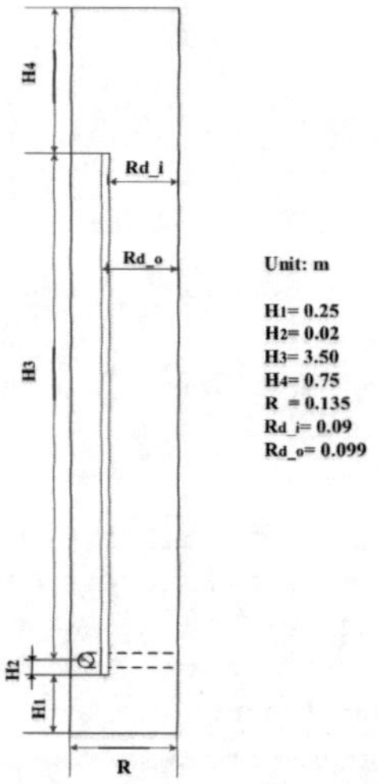

Figure 7. Dimension of the scale-up AR_ALR.

The The overall gas holdup simulated with the larger scale AR_ALR is higher than that of the laboratory scale, as shown in Figure 8. As pointed out by van Baten *et al.* [23], the friction losses of liquid phase encountered in larger scale reactor are reduced and the liquid circulation velocity increases. For AR_ALR, much more bubbles could be entrained into the downcomer and therefore high gas holdup is predicted.

In order to validate the developed CFD model, further simulations with the larger scale reactor were performed on two more high gas superficial velocities and compared with the experimental data from [24]. As shown in Figure 8, somewhat higher simulated gas holdup is, however, obtained probably due to absence of the enlarged gas separator part in the simulated AR_ALR.

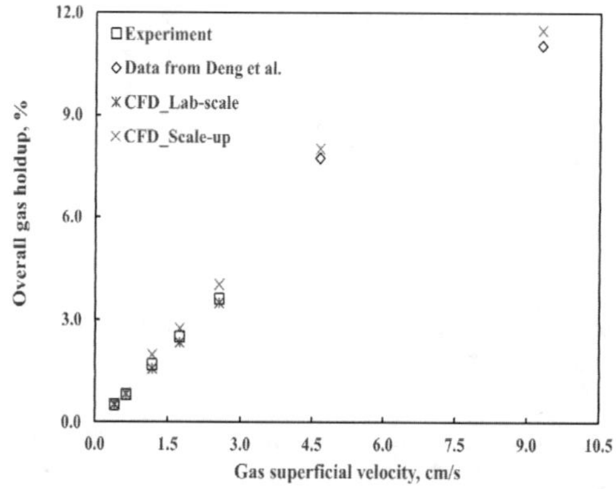

Figure 8. Effect of reactor scale on overall gas holdup: square-experiments; diamond-from [24]; star-CFD with laboratory scale AR_ALR; cross-scale-up AR_ALR.

In Figure 9, the volumetric mass transfer coefficients obtained in simulations and experiments are compared in the two reactor scales. The simulated results with the larger scale reactor are higher than that of the laboratory scale reactor, which is in accordance with the predicted gas holdup.

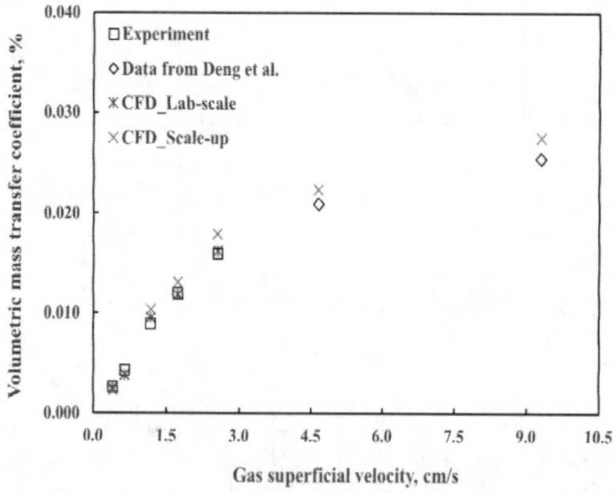

Figure 9. Effect of reactor scale on volumetric mass transfer coefficient: square-experiments; diamond-from [24]; star-CFD with laboratory scale AR_ALR; cross-scale-up AR_ALR.

5 Conclusion

Hydrodynamic characteristic and mass transfer performance of the annulus-rising airlift reactor were investigated with experimental and CFD simulation methods. The effects of superficial gas velocity and reactor scale are presented and compared with experimental results from literature.

A three-phase Eulerian model was developed for simulating the AR_ALR. Good agreements on gas holdup and volumetric mass coefficient were obtained over the range of the studied gas superficial velocities. The three flow regimes observed in experiments in AR_ALR are also captured by the developed model. According to the comparison of experiments and simulations, the gas holdup and the volumetric mass transfer coefficients can be predicted well at different scales and gas superficial velocities using the three-phase Eulerian model with the proposed momentum terms.

References

1. M. Y. Chisti, M. Moo-Young, Chem. Eng. Comm. **60**, 47 (1987)
2. R. Bannari, A. Bannari, P. Vermette, P. Proulx, Biotech. Bioeng. **109**, 8 (2012)
3. M. Y. Chisti, Airlift Bioreactors, London: Elsevier, 8-9 (1989)
4. C. C. Fu, L. S. Fan, W. T. Wu, Chem. Eng. Tech., **30**, 5 (2007)
5. L. C. Shing, H. S. Jye, Chem. Eng. J. **91**, 19 (2003)
6. A. Fadavi, Y. Chisti, Chem. Eng. J. **131**, 6 (2007)
7. H. P. Luo, M. H. Al-Dahhan, Chem. Eng. Sci., **63**, 11 (2008)
8. P. M. Kilonzo, A. Margaritis, M. A. Bergougnou, J.T. Yu, Q. Ye, Biochem. Eng. J., **34**, 9 (2007)
9. Q. S. Huang, C. Yang, G. Z. Yu, Z. S. Mao, Chem. Eng. Tech., **30**, 9 (2007)
10. M. Šimčík, A. Mota, M. C. Ruzicka, A. Vicente, J. Teixeira, Chem. Eng. Sci., **66**, 11 (2011)
11. Q. S. Huang, C. Yang, G. Z. Yu, Z. S. Mao, Chem. Eng. Tech., **31**, 8 (2008)
12. A. Sokolichin, G. Eigenberger, Chem. Eng. Sci., **49**, 11 (1994)
13. Q. S. Huang, C. Yang, G. Z. Yu, Z. S. Mao, Chem. Eng. Sci., **65**, 9 (2010)
14. H. P. Luo, M. H. Al-Dahhan, Chem. Eng. Sci., **30**, 15 (2011)
15. E. R. Gouveia, C. O. Hokka, A. C. Badino-Jr, Brazil. J. of Chem. Eng., **20**, 9 (2003)
16. S. L. Sun, C. J. Liu, W. S. Wei, X. J. Bao, Powder. Tech., **162**, 6 (2006)
17. K. Koide, K. Kurematsu, S. Iwamoto, Y. Iwata, K. Horibe, J. Chem. Eng. Jap., **16**, 6 (1983)
18. K. Koide, K. Kurematsu, S. Iwamoto, Y. Iwata, K. Horibe, J. Chem. Eng. Jap., **16**, 6 (1983)
19. P. R. Lennartsson, C. Niklasson, M. J. Taherzadeh, Bioresour. Tech., **102**, 7 (2011)
20. C. O. Vandu, R. Krishna, Chem. Eng. Process, **43**, 8 (2004)
21. H. M. Letzel, J. C. Schouten, R. Krishna, C. M. van den Bleek, Chem. Eng. Sci., **54**, 9 (1999)
22. A. Cockx., Z. D. Quang, J. M. Audic, A. Liné, M. Roustan, "Global and local mass transfer coefficients in the waste water treatment process by computational fluid dynamics," Chem. Eng. Process, **40**, 7 (2001)
23. J. M. van Baten, J. Ellenberger, R. Krishna, "Hydrodynamics of internal air-lift reactors: experiments versus CFD simulations, Chem. Eng. Process, **42**, 9 (2003)
24. Z. H. Deng, T. F. Wang, N. Zhang, Z. W. Wang, Chem. Eng. J. **160**, 8 (2010)

MATEC Web of Conferences **49**, 04002 (2016)
DOI: 10.1051/matecconf/20164904002

Assessment of Interaction between Baffles and Impeller Blades in Stirred Tanks

Dmitry Gradov [a], Arto Laari and Tuomas Koiranen

Lappeenranta University of Technology, Finland

Abstract. In this work, the sensitivity of hydrodynamics on stirrer position is considered when the Multiple Reference Frame technique is applied to simulate mixing tanks of single and multiphase systems. A typical stirred reactor equipped with four baffles and a six blade Rushton turbine was chosen for the studies. Operational conditions of constant speed (5 rps) and $2.653 \cdot 10^{-4}$ m/s gas superficial velocity with single bubble size (1 mm) were set. The reactor was simulated in Fluent 15 with four different impeller blade positions. The results were averaged and validated against experimental data produced by Particle Image Velocimetry. The comparison was made using profiles of velocity components along the sampled lines representing various zones of the vessel. The impeller angle of rotation with respect to baffle position was found to contribute to the overall reactor hydrodynamics especially for multiphase systems. Averaging of the simulation results was found to improve the accuracy of the simulated mixing hydrodynamics.

1 Introduction

Stirred tank is one of the key unit operations widely utilized in chemical industry. Its performance has direct impact on production quality, yield, and therefore overall effectiveness. That is why it is extremely important to match a certain chemical reaction to a suitable reactor and to operate it under optimal conditions. It is well known that chemical process features and performance can be forecasted and its output estimated via mathematic modelling.

Chemical reaction modelling often employs the assumption of media to be perfectly mixed. In order to assess hydrodynamics within a reactor it is possible to take into account different geometries and scales of operation using Computational Fluid Dynamics (CFD). Reynolds-averaged Navier-Stokes (RANS) equations usage is a compromise between purposes of engineering applications and computational time and resources available.

Since 80's stirred tank modelling has been enlightened in literature by many research groups such as Issa and Gosman, Harvey and Greaves, Lane et al., Vivek et al., Laakkonen et al., Bai et al., and Huang et al among many others [1–7]. Liquid mixing is described numerically mainly via two-equation models of RANS family due to affordable computational load. For that reason, Large Eddy Simulation approach has fewer applications in multiphase mixing regardless the fact that it shows higher accuracy compared to experimental data. Aubin et al. [8] and Vlachakis [9] claim that the impeller interaction with baffles has a periodic character that allows to assume the stirred reactor to be in a steady state. Two main approaches to treat the impeller geometry are generally used when stirred tank is modeled. These are the Sliding Mesh (SM) and the Multiple Reference Frame (MRF) approaches. The MRF simulation aids in saving time without losses in accuracy compared to transient SM approach where the actual geometry of agitator is moving. Unfortunately, despite the warnings stated by Marshal and Bakker in "Computational Fluid Mixing" [10] that mixing tank simulation with frozen impeller (MRF) of outermost positions and averaging the results should be done to get accurate results, this suggestion has not been usually followed in the published literature. Therefore, the influence of impeller blade position on the overall hydrodynamics in a vessel equipped with four baffles is studied in this work and evaluated qualitatively and numerically by averaging the simulation data.

There are several experimental methods for validating numerical simulation data of hydrodynamics. They could be grouped as intrusive methods, which are mainly based on Constant Temperature Anemometry, for example Hot Wire Anemometry (HWA) [11], and non-intrusive ones, such as Laser Doppler Velocimetry (LDA) [12] and Particle Image Velocimetry (PIV) [13]. The PIV technique provides velocity vector fields in two-dimensional space which is used in this study.

The present paper is dedicated to baffle-impeller interaction assessment and results averaging procedure for stirred reactor when the Multiple Reference Frame method is applied. The single and multiphase simulation results of averaging are validated against experimental data obtained by means of PIV.

[a] Corresponding author: dmitry.gradov@lut.fi

2 Materials and methods

2.1 Object of study

A round bottom stirred reactor equipped with baffles and a Rushton turbine is the object of study in the current research. The dimensions of the reactor are presented in Table 1. Made out of glass, the vessel is sealed in plexiglass box for conducting PIV tests. The clearance of the six-blade impeller was kept the same as its diameter. The gas inlet was mounted in the very center of the vessel bottom. The operational conditions were 300 rpm of rotational speed and half a liter per minute for gas flow rate. In order to examine the influence of the blades positioning far from the impeller, the liquid height was twice the tank diameter. The schematic view of the reactor together with velocity profile sampling strategy is illustrated in Figure 1.

Table 1. Dimensions of the reactor.

Reactor Dimension	Value, mm
Tank diameter	20
Liquid height	40
Impeller diameter	70
Impeller blade width	15
Impeller clearance	70
Shaft diameter	14
Baffle width	20
Baffle distance from tank wall	5
Sparger height	10
Sparger width	7

2.2 Validation technique

Particle Image Velocimetry is widely used for fluid flow study and its results are frequently used to validate simulation results. It produces instantaneous velocity vector field in 2D or 3D in case of tomographic PIV. By means of fast laser pulsing it is possible to track slight shift of submerged light tracers within an interrogation window of interest. A classical set-up includes cameras installed perpendicular toward dense laser sheet produced by Nd:Yag laser. In the current research, two CCD cameras with a resolution of 1600 pixels × 1200 pixels were used. That allows capturing 2D velocity fields from the whole reactor with the resolution of 188 vectors × 189 vectors. The main requirements for the tracer particles are that they follow the media without disturbing the flow and reflect intensive light on certain wavelengths. It is also necessary to take into account that the studied reactor has many strongly reflecting parts which prevents the use of standard hollow glass PIV tracer particles. Therefore, 20-50 μm particles made of melamine resin covered with fluorescent dye (Rhodamine B) were applied. Their usage in water flow tests has the benefit that more uniform dark background can be obtained with large camera aperture installing wave length filters on the cameras. To detect the movement of gas phase, tests were carried out with only grey filters mounted on the cameras to decrease the reflected light intensity from bubbles to protect the camera.

The position of the laser sheet was set to be at a distance of 10 mm in front of a baffle to maximize the area captured in a frame and to eliminate the neighboring baffle from the field of view. Each frame includes half of the reactor from the vessel wall to the agitator shaft in width and 10 cm in height. Therefore, six overlapping frames were taken along the vertical axes in order to capture the whole tank. The coordinates of the interrogation window were measured to extract the analogue plane from CFD simulation results. A schematic representation of the PIV set-up is shown in Figure 2

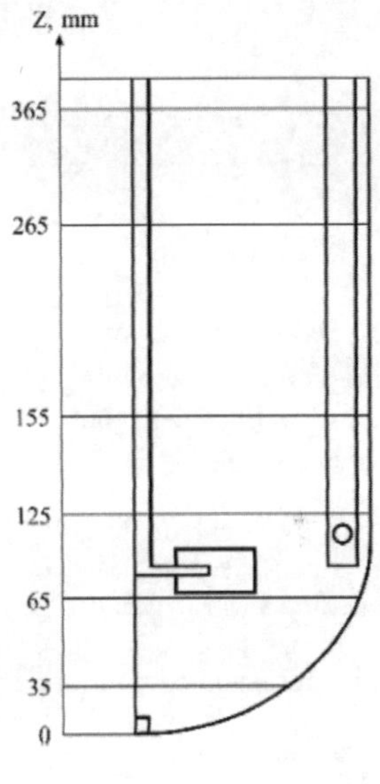

Figure 1. Mixing reactor and schematic representation of profiles sampling.

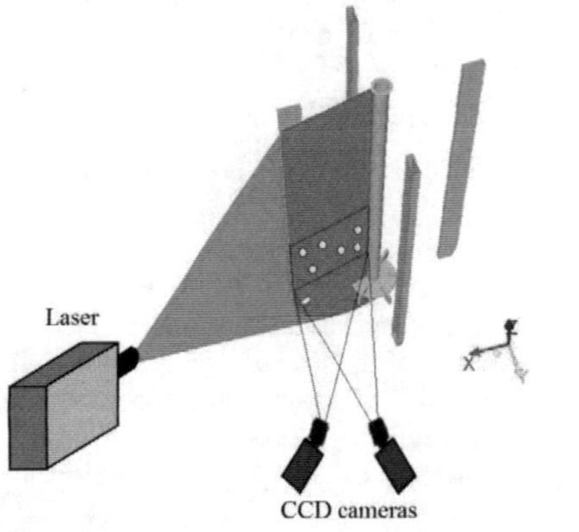

Figure 2. Conceptual schema of stereo PIV set-up used in the current work.

The detention time between the laser pulses was adjusted individually for each position along the vertical axis so that tracer shift would not exceed the interrogation area when post-processing the data. In total, 500 frames were taken to obtain experimental results at steady state.

2.3 Simulation settings

It was decided not to use symmetry and have full reactor hydrodynamics so that there would be no reasonable questions raised on how symmetry may affect the overall flow field. Reactor domain has been split by structured meshes with maximum aspect ratio 30 and high orthogonal quality to reduce numerical errors. Grid independence test conveyed over three meshes with increasing resolution showed that it has no sense to go more than 800000 elements, as flow does not change significantly. Normalized root mean square error index (NRMSE) (eq. 1) was under 10 %, which identifies grid difference to be insignificant [14].

$$nrmse = \frac{3}{(n_{fine}/n_{coarse})^p - 1}\sqrt{\frac{\sum_{i=1}^{N}(\phi_1(i)-\phi_1(i))^2}{\sum_{i=1}^{n}(\phi_2)^2}} \cdot 100\% \quad (1)$$

n_{fine} and n_{coarse} are the number of cells in a fine and coarse grids correspondingly, p represents the order of discretization, N is the number of samples, ϕ_1 and ϕ_2 are the predictions on coarse and fine grids accordingly.

Multiphase mixing was simulated in Eulerian-Eulerian model according to which the phases may interpenetrate each other. Five impeller revolutions per second produce fully turbulent flow regime (24000 Re), therefore the k-ε Realizable model was used for turbulence modelling since it was designed to overcome the difficulties of the Standard k-ε model when rotation and/or swirling are involved via the introduction of turbulence viscosity as a variable instead of being constant [15]. The impeller motion was modelled via the MRF approach where the fluid in the region surrounding the rotating parts moves around a frozen impeller and a shaft. By producing boundary conditions for the interfacing walls between the zones, the motion transfers to the outer stationary zone.

To avoid computational instabilities, mixing of the liquid phase only was simulated first until convergence was reached and then the gas flow rate was activated with a constant bubble size of 1 mm. The reactor top was assigned as pressure outlet with backflow volume fraction of air set to zero, which corresponds to degassing condition. The interphase interaction included only the drag force since the cumulative effect of lift, wall lubrication and turbulence diffusion forces can be considered negligible according to Lane et al. and Scargiali et al. [16, 17]. Scargiali *et al.* and Brucato et al. [18] showed superior performance of the Morsi-Alexander drag force in combination with the Brucato modification coefficient included. Originally, the Brucato coefficient was suggested for solid particles being later reduced by two orders of magnitude to adopt it for gas bubbles [18, 19].

Third order differencing QUICK (Quadratic Upwind Interpolation for Convective Kinetics) scheme was used for momentum and turbulence because of its higher accuracy. Coupling of pressure and velocity was performed by using Semi-Implicit algorithm. The simulations were continued until the residuals of all the variables were lower than 10^{-4}. Two-phase simulations were run until both the convergence and the error in the air mass balance between the inlet and the outlet reached satisfactory levels.

Single phase mixing computation took around 8500 iterations, which corresponds to 14 hours of real time when four cores were involved (Intel(R) Xeon(R) w3565 3.2 GHz and 24 Gb RAM) Multiphase simulations took no more than 18000 iteration and 30 hours of real time.

2.4 Averaging procedure

The reactor construction and the four baffles against the six blades of Rushton turbine allow positioning the impeller within angles between 0° to 60° with respect to the baffles. The simulations were done at four positions of the impeller 0°, 15°, 30°, and 45° correspondingly, as it is depictured in Fig. 3. Then, the velocity profiles representing the different regions of the reactor were sampled from PIV and CFD results and averaged in Matlab over the different combinations of the impeller positions. The resulting data of the single phase and multiphase mixing have been plotted versus the PIV tests. The major complication in the averaging is to find the data correlation when the number and the shape of the elements differ as the impeller turns. In this work, all the data correlation was accomplished manually. However, it is recommended that in future automatic CFD tools for averaging should be created.

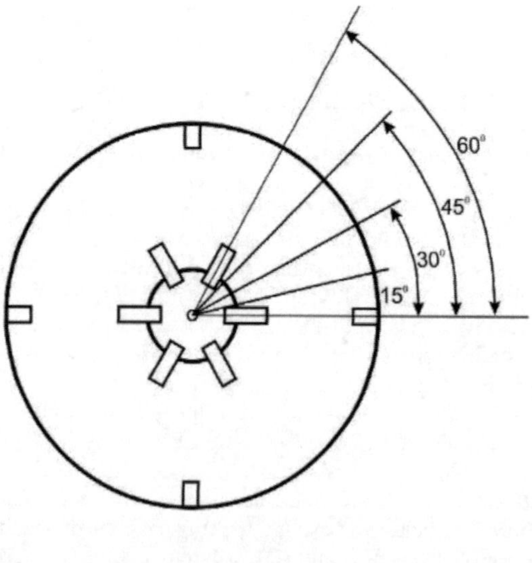

Figure 3. Impeller positioning diagram.

3 Results

3.1 Single phase averaging assessment

Before the averaging results are tested, it is interesting to see the velocity component profiles at different positions along the reactor height. The vessel may be split into three regions as follows: the bottom, impeller and reactor upper part regions. The profiles have been chosen so that all the three parts are tested. The results are gathered in Figures 4 and 5.

The major difference in the velocity components can be seen at the 35, 65 and 365 mm profiles whereas the 125, 155 and 265 mm ones indicate no dispersion. The rotating zone expectedly produces the highest velocity variation as how in the 125 mm profile. The MRF technique employs averaging at the interface walls before the motion is transferred to the stationary zone. By careful examination of the variation of the velocity components along the sampled profiles, it can be noticed that the biggest deviation is between the angles 0° and 30°, which signify that these two are the extreme impeller positions. In addition, some deviation may be noticed over the distance between the bottom and the top parts of the tank.

The RANS models are known to produce accurate results near the source of energy and in the profiles at 65, 125, 155 mm the simulation results are qualitatively in the best agreement with the experimental ones. The RANS models have difficulties with flow separation and swirling predictions and they are greatly dependent on the domain geometry and the grid configuration. The round bottom and the position of the baffles create difficulties for prediction of hydrodynamics in the bottom part, as represented by the 35 mm profile. In the regions that are far from the source of energy, the agreement is very poor.

It is important to know the sufficient number of samples that should be used when averaging the data. Therefore, it was decided to use the two extreme impeller positions, for example 0° and 30°, and all four samples for testing purposes. The outputs are presented in Figure 6.

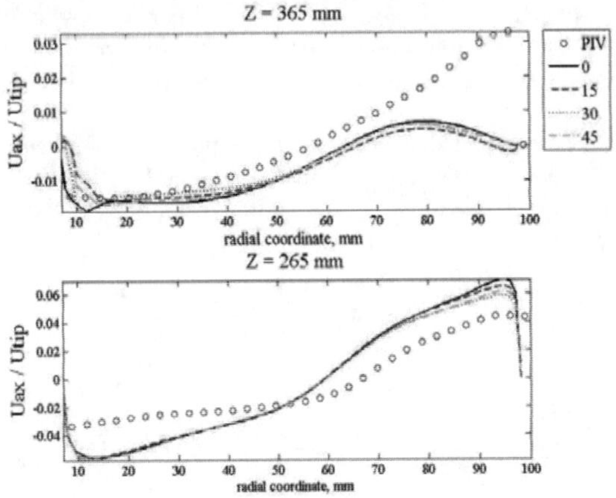

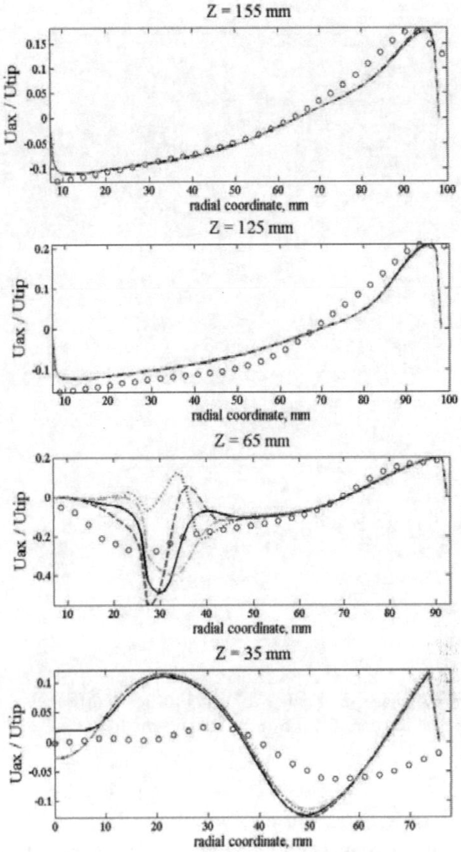

Figure 4. Comparison of axial velocity profiles at different impeller positions produced by averaged results from CFD and PIV

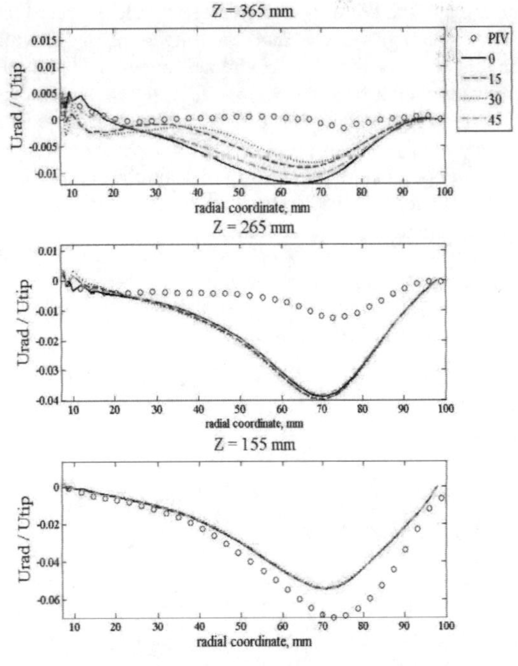

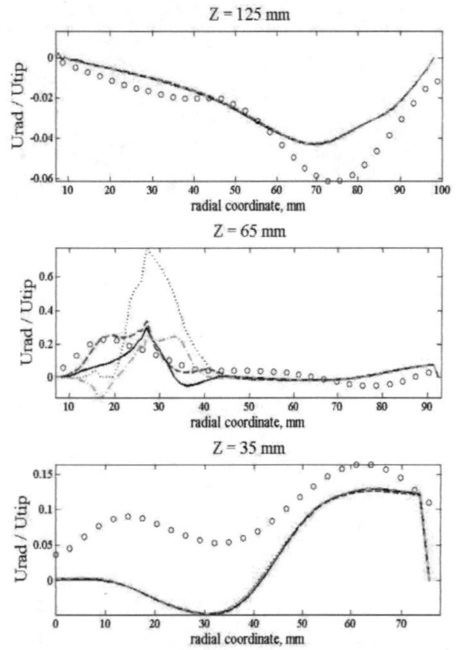

Figure 5. Comparison of radial velocity profiles at different impeller positions produced by averaged results from CFD and PIV

The biggest deviation can be observed in the close vicinity of the impeller in the rotating zone, which is shown in the 65 mm profile. Therefore, it is recommended to do the averaging of the results over all the four angles of the agitator positions when this particular zone is crucial in the research. What comes to the top and the bottom parts, averaging over all positions is not that necessary and in single phase mixing simulation of the two extreme impeller locations produces similar results.

Quantitative evaluation of improvement achieved by averaging of results is of great interest and may also give valuable information about how the number of samples taken into account may improve the simulation results in comparison to experimental data.

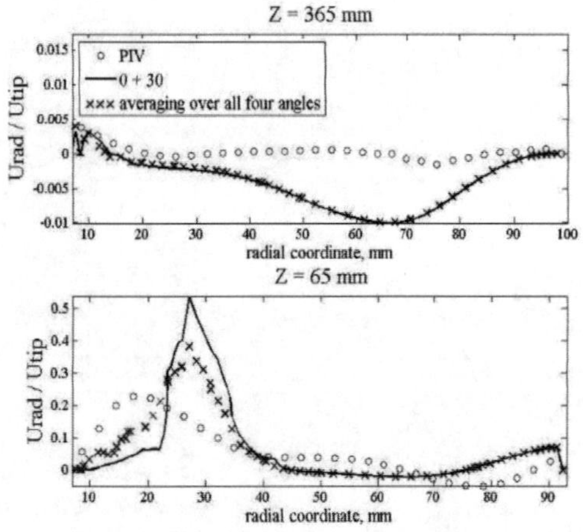

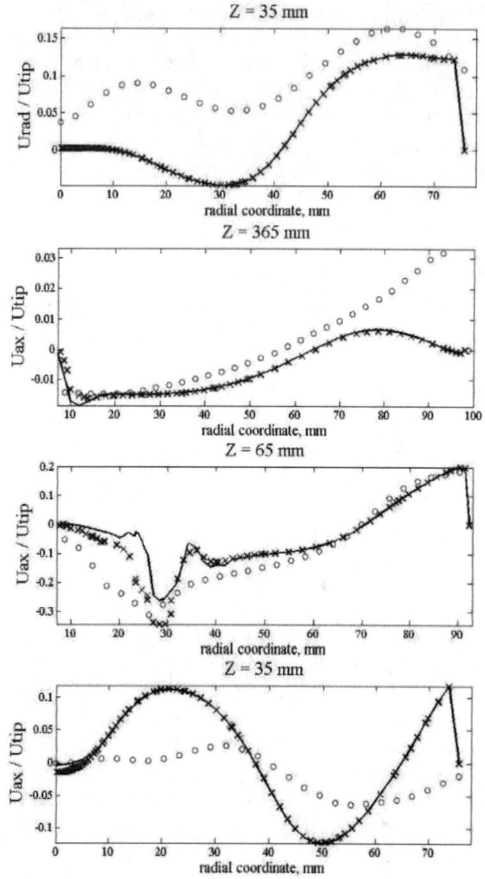

Figure 6. Axial and radial velocity profiles comparison of the CFD simulations averaged over different impeller positions and the PIV results

3.2 Multiphase averaging assessment

Importance of the averaging procedure represents particular interest for multiphase systems. Gas-liquid mixing was considered next and the influence of the impeller position relative to baffles was studied by comparing the velocity profiles at different locations. It must be noted that multiphase mixing is a complicated phenomenon that includes multiple counter acting forces and the simulation of its hydrodynamics has greater uncertainty when compared to experimental results than that of the single phase.

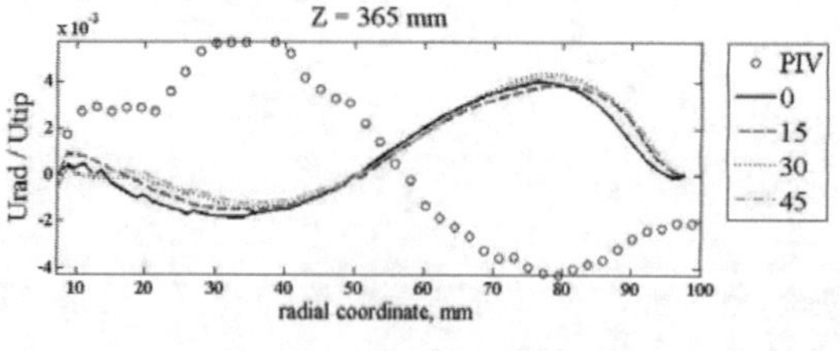

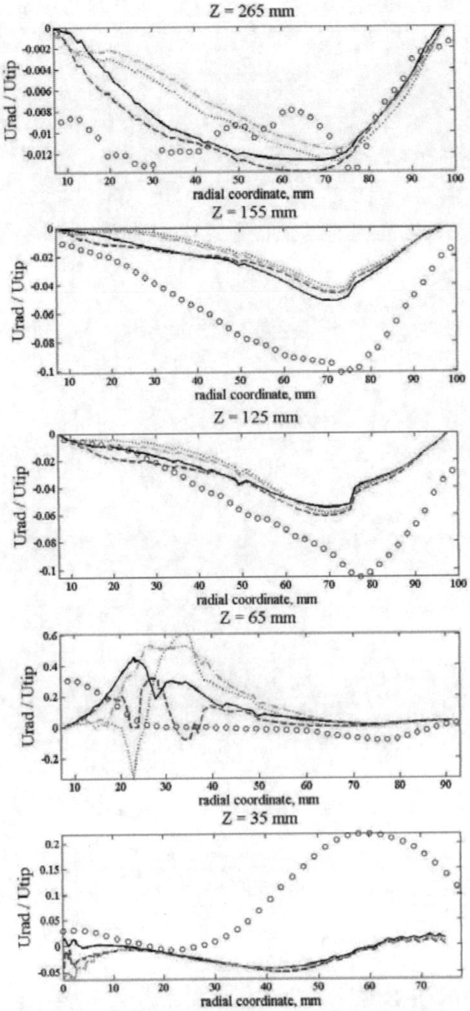

Figure 7. Radial velocity profiles comparison of the CFD simulations in liquid phase averaged over different impeller positions and the PIV results

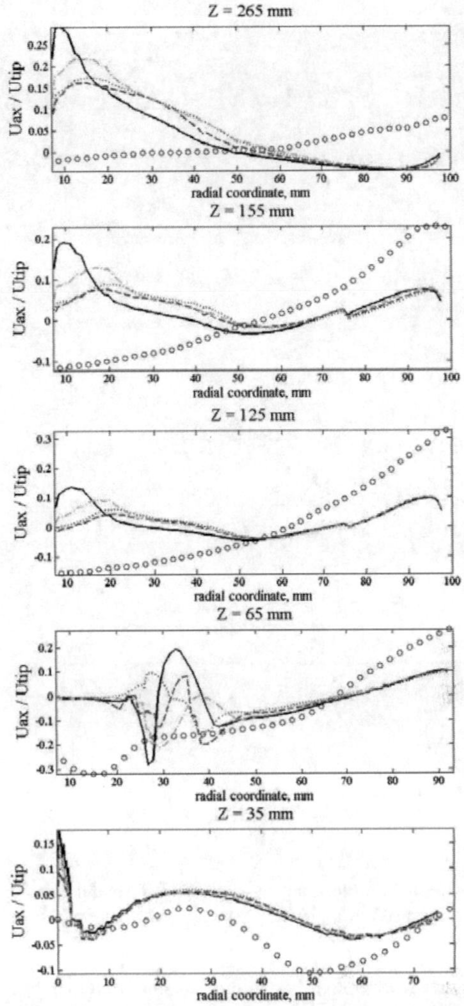

Figure 8. Axial velocity profiles comparison of the CFD simulations in liquid phase averaged over different impeller positions and the PIV results

The profiles that are shown below in Figs. 7 - 10 represent the complete reactor from the top to bottom and from the vessel center, or agitator shaft, to the tank wall. Such a lay-out helps the reader in imagining the whole reactor when looking at the profiles and assessing qualitatively the hydrodynamics in the reactor. Since the multiphase system is complicated with interfacial forces and turbulence computed per phase, even a slight change in the impeller blade position gives deviation in the velocity components in every sampled profile. As in the case of the single phase, the biggest variation is observed near the impeller.

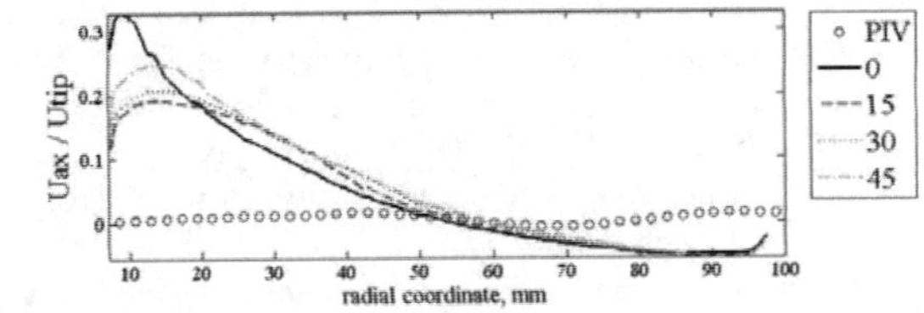

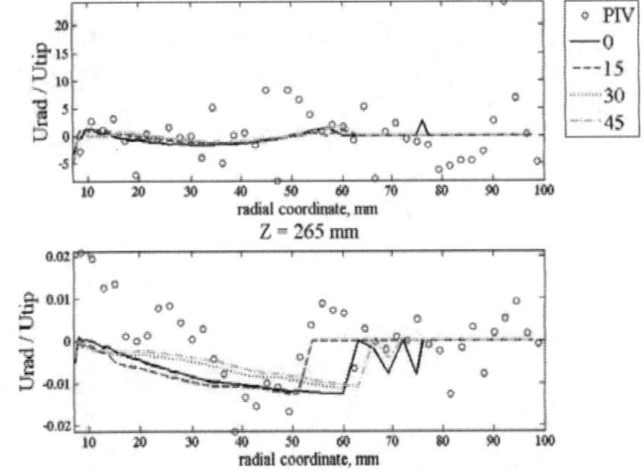

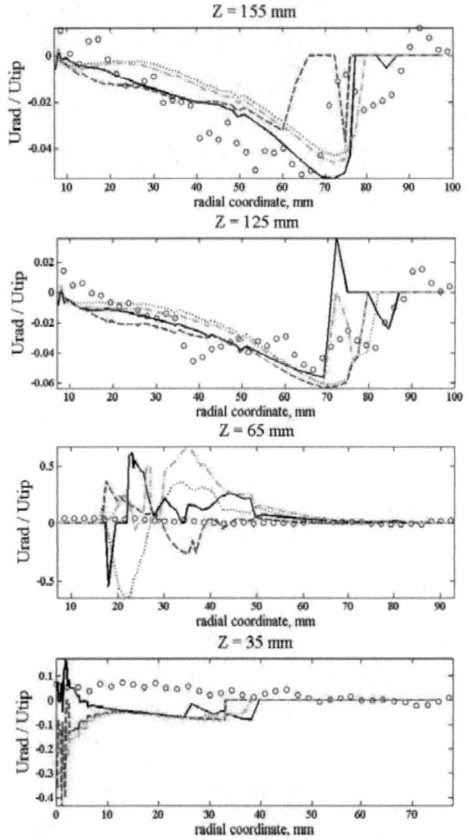

Figure 9. Radial profiles comparison of the CFD simulations in gas phase averaged over different impeller positions and the PIV results

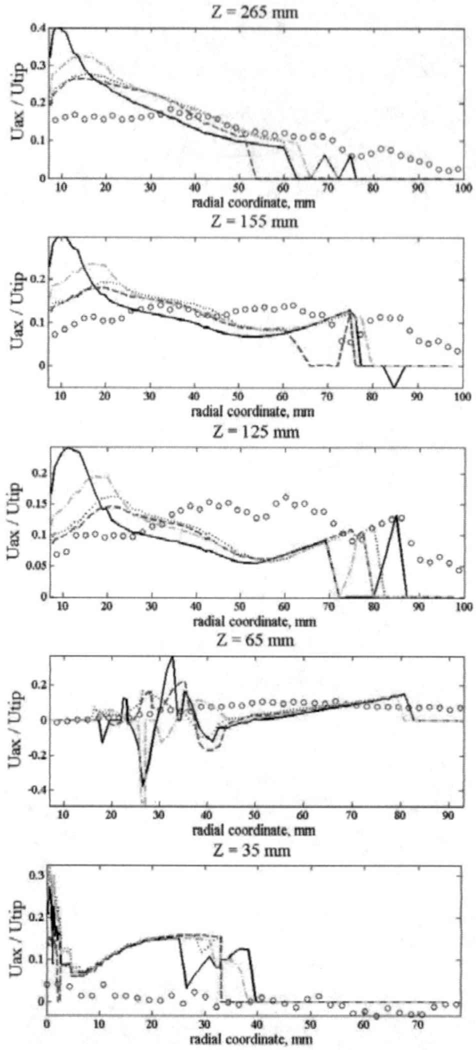

Figure 10. Axial velocity profiles comparison of the CFD simulations in gas phase averaged over different impeller positions and the PIV results

One may note that the angle of the agitator with respect to baffles has a bigger effect on the gaseous phase. In the current work the gas flow rate is rather low and therefore, the primary phase turbulence has great impact on the movement of the secondary phase. Thus, it is important to undertake averaging in order to produce results of higher accuracy. Also, it could be expected that at higher gas flow rates gas hydrodynamics should be less dependent on the fluctuations in the liquid flow and therefore, averaging over the two extreme positions for gas phase should be enough. However, the number of impeller positions to be involved in the averaging is dictated by the qualitative evaluation of hydrodynamics.

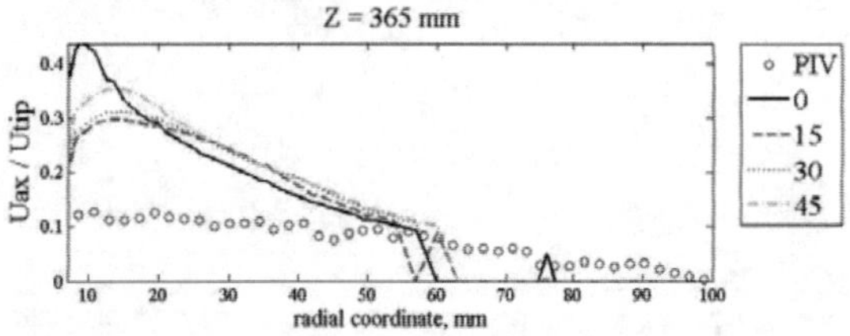

4 Conclusions

The results in the presented work clearly indicate that averaging procedure may improve CFD simulation results of single and multiphase mixing in a baffled stirred tank when "frozen rotor" approach is applied. It was observed that the profiles of the velocity components depending on the impeller angle relative to the baffles. Such variation is low for single phase and higher significant when a secondary gaseous phase is introduced. It was found that averaging over the two extreme impeller positions improves the hydrodynamics results. As far as the Rushton turbine is concerned, averaging over four impeller positions produces more accurate results than just over two extreme positions of the stirrer. Those researchers and engineers who deal with the topic of the paper are advised to execute several simulations with the stirrer turned within the available sector. Such a sector is dictated by the number of impeller blades relative to the quantity and construction of baffles. Since the frozen rotor technique is widely used in mixing simulations, it makes sense to consider developing automatic or semiautomatic methods for averaging procedure.

References

1. V. R. Vivek, Chem. Eng. Commun., **189**, 895-922 (2002)
2. B. Bai, Y. Wang, P. M. Armenante, Int. J. Pharm., **403**, 1-14 (2010)
3. W. Huang, K. Li, Nuclear Reactor Therm. Hydr. and Other Applications, (2013)
4. R. I. Issa, A. D. Gosman, 2nd International Conference of Numerical Methods in Laminar and Turbulent Flow (1981)
5. P. S. Harvey, M. Greaves, Transactions of the Institution of Chemical Engineers, **60**, 195-200 (1982)
6. G. L. Lane, M. P. Schwarz, G. M. Evans, 10th European Conference on Mixing, 197-204 (2000)
7. M. Laakkonen, P. Moilanen, V. Alopaeus and J. Aittamaa, Chem. Eng. J. **109**, 37-47 (2005)
8. J. Aubin, S. M. Kresta, J. Bertrand, C. Xuere, D. F. Fletcher, Transactions of Institution of Chemical Engineers, **84**, 569-582 (2006)
9. V. N. Vlachakis, M.S. Thesis in Mech. Eng. (2006)
10. M. Marshall, A. Bakker, Computational Fluid Mixing (2002)
11. K. D. Jensen, J. Braz. Soc. Mech. Sci. & Eng., **26**, 400-419 (2004)
12. H. Wu, G. K. Patterson Chem. Eng. Sci., **44**, 2207-2221 (1989)
13. L. Xinhong, B. Yuyun, L. I. Zhipeng, G. Zhengming, Chin. J. Chem. Eng., **18**, 588-599 (2010)
14. H. Wang, Z. Zhai, Build. Env. J., **52**, 107-118 (2012)
15. T. H. Shih, W. W. Liou, A. Shabbir, Z. Yang, J. Zhu, Comput. Fluids, **24**, 227-238 (1995)
16. G. L. Lane, M. P. Schwarz, G. M. Evans, 10th European Conference on Mixing, 197-204 (2000)
17. F. Scargiali, A. D'Orazio, F. Grisafi and A. Brucato, Chem. Eng. Res. Des., **85**, 647-653 (2007)
18. A. Brucato, F. Grisafi, G. Montante, Chem. Eng. Sci., **53**, 3295-3314 (1998)
19. M. Laakkonen, PhD Thesis in Tech. (2006)

MATEC Web of Conferences **49**, 04003 (2016)
DOI: 10.1051/matecconf/20164904003

RO Concentrate Softening through Induced Crystallization in a Fluidized Bed Reactor

Kecheng Liu [1], Qun Huang [2], Xiao Long [1], Ziyue Cao [2], Baiqing Zhou [2, a]

[1] *Hebei Electric Power Commissioning Institute, Shijiazhuang 050021, PR China*
[2] *School of Power and Mechanical Engineering, Wuhan University, Wuhan 430072, PR China*

Abstract. Induced crystallization technique was adopted in a fluidized bed reactor to remove scale-forming ions from RO concentrate in this paper, the result of which is as follows: the effluent of the fluidized bed reactor reached stability in two minutes after the operation started; the optimized consumption ratio of softener was 1, under the circumstances of which the removal rate of Ca^{2+} was 89.22%; the removal of Ca^{2+} was mainly completed at the position 20 centimeters above the bottom of the bed; the reflux rate had no influence on the removal of Ca^{2+}, but it had influence on the turbidity of effluent.

1 Introduction

Treating water with RO (Reverse Osmosis) process will produce RO concentrate. Especially when using RO to produce fresh water from brackish water, the concentrate yield may be higher than the fresh water yield [1]. At present, the total output of seawater desalination in China is 5×105m3/d [2], while the recovery rate of RO when treating seawater is 30% to 40% [3]. That is to say, the daily fresh water output is 7.5×105m3～1.17×106m3. Therefore, if RO concentrate is directly discharged without any recycling method, it would not only be a waste of water resources, but also affect the environment.

So far, the treatments of RO concentrate mainly include deep-well injection [4, 5], flush filters [6], boiler dedusting [7], salt manufacturing [8], electrolysis [9, 10], membrane distillation [11] and membrane crystallization [12]. The author proposed the method of softening RO concentrate through induced crystallization in a fluidized bed reactor and conducted relative experiments.

Induced crystallization means making oversaturated $CaCO_3$ and $CaSO_4$ grow on the added seed crystal to achieve the goal of reducing the concentration of scale-forming ions in the solution. Induced crystallization coupling with a fluidized bed reactor means using fluidized bed packing as seed crystal and remove scale-forming ions during the process of oversaturated RO concentrate flowing from bottom to top in the fluidized bed reactor.

The crystallization process depends on the equilibrium relation between the solid and the solution. If the solution is already supersaturated when solid substance comes into contact with it, then the supersaturated portion of the substance in the solution will crystallize [13]. The crystallization process of RO concentrate can be divided into two parts, one is the formation of crystal nucleus and the other is the growth of crystals, which is the growing process of crystal nucleus [14]. And the mechanism of induced crystallization lies in adding crystal nucleus. Theoretically, induced crystallization can remove every substance that can precipitate as crystals from water solution [15].

Table 1. Experimental water quality

DD	pH	LSI	JD_m	YD	$1/2Ca^{2+}$	$1/2Mg^{2+}$	Cl^-	SO_4^{2-}
3587	7.89	1.813	21.84	36.90	27.95	8.954	350.64	308.8
Ba^{2+}	Sr^{2+}	SiO_2	TDS	K^+	Na^+	Fe^{3+}	F^-	
0.2210	1.690	39.65	2661	3.100	98.52	2.140	1.587	
Unit of measurement: DD: μS/cm; pH and LSI have no unit; JD_m, YD, $1/2Ca^{2+}$ and $1/2Mg^{2+}$: mmol/L; other indicators: mg/L.								

[a] Corresponding author: bqzhou@whu.edu.cn

2 Material and method

2.1 Experimental water

Prepare the experimental water based on the average water quality (as shown in Table 1) of the concentrate of two single-pass two-stage RO systems in Unit 3 and 4 of a power plant. The scale inhibitor (PTP-0100) content of the experimental water is 16mg/L.

2.2 Experimental facility

The experimental facility and process consists of a drug tank, a raw water tank, a dosing pump, a feed water pump, a fluidized bed, four sample connections, three flowmeters and other devices, as shown in Figure 1.

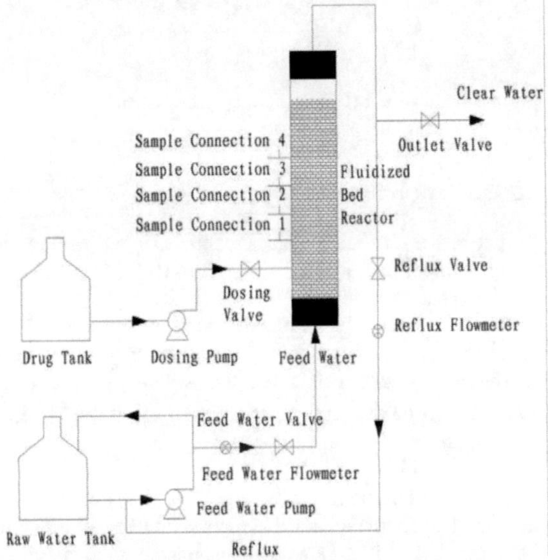

Figure 1. Process of induced crystallization in a fluidized bed reactor

The fluidized bed reactor is made of PMMA with the height of 100 centimeters and the inner diameter of 4 centimeters. A screen is installed inside the fluidized bed reactor to prevent the inducer from flowing out. The drug injection tube is introduced into the reactor 2 centimeters above the screen. Sample connections are installed 20 centimeters, 40 centimeters, 60 centimeters and 80 centimeters above the bottom of the reactor respectively. Limestone is used as fluidized bed packing, and the static bed height is 40 centimeters. Feed water flows into the reactor at the bottom of the bed and flows out from the outlet at the top of the reactor. The packing acts not only as the inducer but also as the sludge blanket in a clarification tank, which bear the functions of induced crystallization, clarification and filtration.

2.3 Experimental method

The operation process of the experimental facility is as follows: The experiment is conducted under room temperature, and each experiment uses brand-new inducers. Turn on the feed water pump to start the experiment. Open the inlet and outlet valves and adjust the inlet flow to the set value. Test the turbidity of outlet water every 30 seconds. After the turbidity of outlet water is stabilized, open the dosing valve to add softener into the reactor. Promptly adjust the dosing flow to the set value, and the induced softening clarification process will be started. Then, test the Ca^{2+} of outlet water every 2 minutes, after the value of which is stabilized, test the Ca^{2+}, pH, hardness and turbidity of the outlet water from each sample connection. After that, turn on the reflux valve, and adjust the outlet valve until the reflux flow reaches the set value. Reduce the softener dosage according to the "feed water/concentrate" ratio. Test the Ca^{2+} and hardness of the outlet water every 2 minutes.

2.4 Fluidizing velocity

The necessary and sufficient condition for operating the fluidized bed reactor is: $u \in (u_{min}, u_{max}]$. Here, u, u_{min} and u_{max} respectively represent the flow rate, the minimum fluidizing velocity and the maximum fluidizing velocity of the fluidized bed reactor.

(1) u_{min}

In order to ensure that the bed is fluidized, u must be higher than u_{min}. u_{min} is calculated by Formula 1 to 3 as follows.

Galileo Number (G_a):

$$G_a = \frac{d^3 \rho (\rho_p - \rho) g}{\mu^2} \tag{1}$$

where, d represents packing particle diameter (m); ρ represents water density, (1000kg/m^3); ρ_p represents packing density (kg/m^3); g represents gravitational acceleration, (9.81m/s^2); and μ represents water viscosity (1.46×10^{-3}Pa·s under 7.0 ℃). Through measuring, d is 0.315 to 0.500mm and ρ_p is 2500kg/m^3 in this experiment.

Fluidization Reynolds Number (Re_{mf}):

$$Re_{mf} = \sqrt{33.7^2 + 0.0408 G_a} - 33.7 \tag{2}$$

Minimum Fluidizing Velocity (u_{min}):

$$u_{min} = \frac{Re_{mf}\,\mu}{\rho d} \tag{3}$$

Because it has to be ensured that inducers with the maximum particle diameter be in a state of suspension, d in Formula 1 to 3 must take the value of the maximum packing particle diameter, which is 0.500×10^{-3}m. Based on computation result, u_{min} is 5.44m/h.

(2) u_{max}

To ensure that the packing will not flow out of the reactor, u must not be higher than u_{max} . u_{max} is related with the height of the bed and the reactor, and needs to be determined by experiment. The experiment process is as follows: Turn on the feed water pump after evacuation of all the air inside the fluidized bed reactor. Open the inlet and outlet valves and adjust the inlet flow while observing the expansion height of the bed. When the bed is expanded to a place approximately 5 centimeters below the outlet, record the current inlet flow which is the maximum flow (Q_{max}) allowed in the reactor. Based on the experiment result, Q_{max} is 90L/h. And by calculation, u_{max} should be 71.66m/h.

2.5 Retention time

The retention time of concentrate in the bed (t) is calculated by Formula 4:

$$t = \frac{3.6(\frac{1}{4}\pi\Phi^2 H - \frac{m}{\rho})}{Q} \quad (4)$$

where, t represents the retention time of concentrate in the bed (s); Φ represents the inner diameter of the reactor (cm); H represents the bed height during steady operation (cm); m represents the mass of inducer inside the reactor (g); ρ represents the inducer density (2.50g/cm^3) and Q represents the inlet flow (L/h).

Based on the experiment result, H is 91cm under the condition when Q is 90L/h and m is 660g. Therefore, t is calculated to be 35.12s.

2.6 Softener

Based on the experimental water quality in Table 1, NaOH is selected as the softener. The amount of softener is expressed through the consumption ratio of softener (N), where N is defined as the molar ratio of NaOH to the experimental water alkalinity (JD_m). In order to ensure complete neutralization of NaOH and HCO_3^-, the theory value of N should be 1.0.

3. Results and Discussion

3.1 Induction time

In the light of The Test Method of Water Coagulation and Precipitation Test Cup (GB/T 16881-2008), induced precipitation time (hereinafter referred to as "induction time") is determined under the temperature of 10℃ with softener dosage of 16mg/L. The relation between the residual Ca^{2+} after the experimental water is softened by induced precipitation and the induction time is shown in Figure 2, where the value of N in curve a and c is 0.8, the inducer dosage is 0 and 30g/L, respectively; while the value of N in curve b and d is 1.1, the inducer dosage is 0 and 30g/L, respectively.

During the crystallization process, inducer acts as the crystal seed, lowering down the energy barrier of the reaction. Therefore, under the same circumstance, adding inducer will accelerate the crystallization process and reduce the residual Ca^{2+}. As reflected in Figure 2, curve c is below curve a while curve d is below curve b.

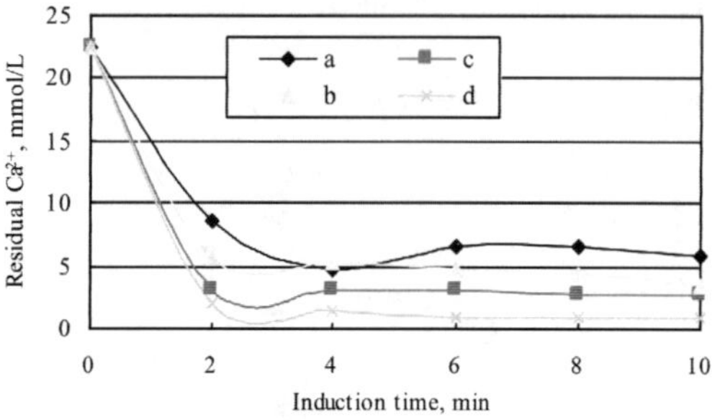

Figure 2. Relation between the residual Ca2+ and the induction time

3.2 Operation time

Reflux rate is defined as the percentage of reflux flow in the total inlet flow.

Using the experimental facility shown in Figure 1, the relation between the removal rate of Ca^{2+} and the operation time is shown in Figure 3, under the condition where the reflux rate is 50%, the water temperature is 19.3℃, u is 63.59m/h, t is 46.38s and N is 1.02. Here, the beginning of the operation is the time when softener starts to be added.

After the softener is added, the CO_3^{2-} level of the inlet flow will keep rising until a supersaturated solution of Ca^{2+} and CO_3^{2-} is formed. The supersaturated solution flows through the fluidized bed reactor and induces the crystallization process. The generated crystals will adhere to the surface of inducer particles, and the Ca^{2+} level will drop. This reaction process takes approximately 2 minutes according to Figure 3.

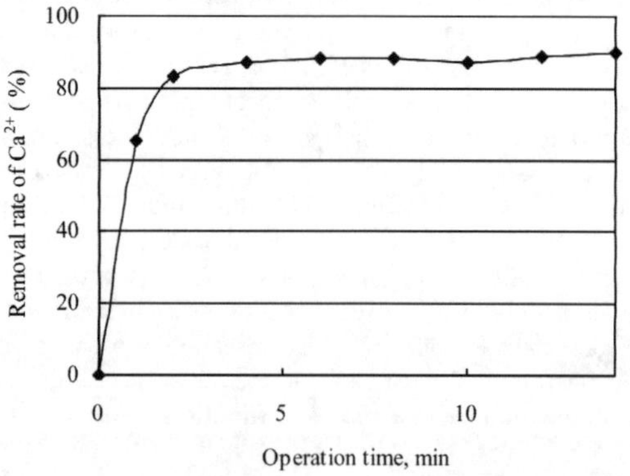

Figure 3. Relation between the removal rate of Ca^{2+} and the operation time

3.3 Softener dosage

Under the same condition as Section 2.2, the relation between the removal rate of Ca^{2+} and N is shown in Figure 4.

When $N < 1.0$, the CO_3^{2-} level rises with the value of N, and the solution goes through the stages of unsaturation, saturation and finally supersaturation. The removal rate of Ca^{2+} increases with the value of N until N reaches 1.0. When $N > 1.0$, there will be no HCO_3^- left in the inlet flow. And keep adding the value of N will no longer affect the CO_3^{2-} level, which means the supersaturated degree of the solution will not change, and the removal rate of Ca^{2+} will maintain at the level when N reaches 1.0.

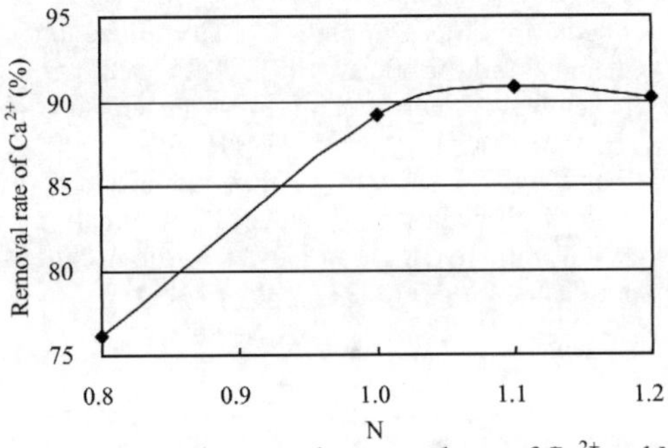

Figure 4. Relation between the removal rate of Ca^{2+} and N

3.4 Height of the fluidized bed reactor

Using the experimental facility shown in Figure 1, the relation between the removal rate of Ca^{2+} and the height of the reactor is shown in Figure 5, under the condition where the reflux rate is 50%, the water temperature is 19.6℃, u is 65.97m/h, t is 46.19s and N is 1.04.

The taller the reactor is, the longer the inlet flow will be in contact with the inducer, and the higher the removal rate of Ca^{2+} will be. However, when the height of the reactor is enough to ensure adequate crystallization time, raising the height will have little effect on the removal rate of Ca^{2+}.

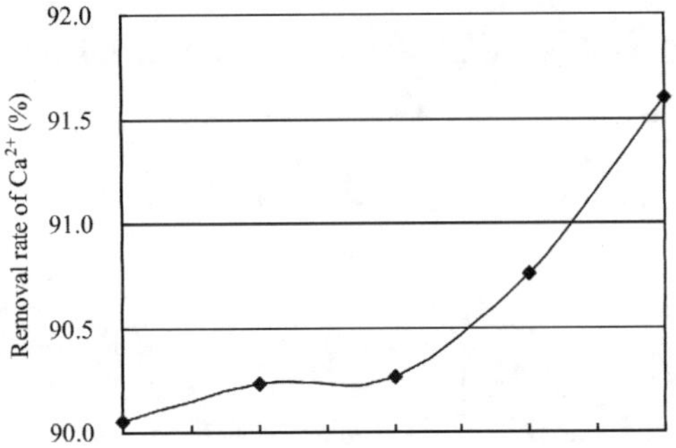

Figure 5. Relation between the removal rate of Ca^{2+} and the height of the reactor

3.5 Reflux rate

Using the experimental facility shown in Figure 1, the relation between the removal rate of Ca^{2+}, the effluent turbidity and the reflux rate is shown in Figure 6, under the condition where the water temperature is 19.6℃, u is 66.35m/h, t is 46.17s and N is 1.02.

In the fluidized bed reactor, the inducer layer is acting as a filter bed. Therefore, higher reflux rate means longer filtering time of the inlet flow and lower effluent turbidity. Nevertheless, if the reflux rate is too high, the pore size of the filter bed will become larger and the filter bed will be less capable of intercepting suspended solids, resulting in higher effluent turbidity.

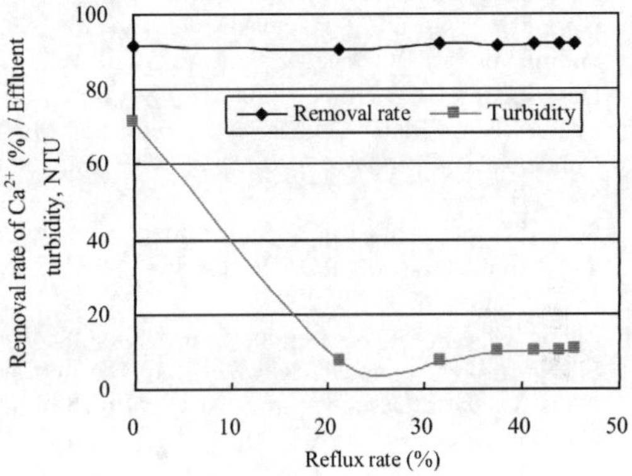

Figure 6. Relation between the removal rate of Ca^{2+} /effluent turbidity and the reflux rate

4. Conclusion

RO concentrate softening through induced crystallization in a fluidized bed reactor has the following characteristics:

(1) The start-up speed of the reactor is relatively fast. The effluent quality can reach stability in approximately 2 minutes after the operation starts, as shown in Figure 2 and 3.

(2) The optimized consumption ratio of softener is 1. When softener is inadequate (N=0.8), the removal rate of Ca^{2+} merely reaches 76.18%. When softener is just enough to completely neutralize HCO_3^- (N=1.0), the removal rate of Ca^{2+} can reach 89.22%. Keep adding softener would not improve the removal rate of Ca^{2+} obviously. As shown in Figure 2 and 4.

(3) The removal of Ca^{2+} was mainly completed at 20 centimeters above the bottom of the bed. For comparison, the removal rate of Ca^{2+} is 90.05% and 91.60% respectively at 20 centimeters and 100 centimeters above the bottom of the bed. The removal rate at the latter position is merely 1.72% higher than that at the former one, as shown in Figure 5.

(4) The reflux rate has no influence on the removal of Ca^{2+}, but it had great influence on effluent turbidity. When the reflux rate is higher than 20%, the effluent turbidity will be lower than 10.5 NTU, as shown in Figure 6.

References

1. Shi Weiping. Discussion and Countermeasure of the Comprehensive Utilization of RO Condensate [J], Energy Saving of Non-Ferrous Metallurgy (China), **21**, 15 (2004)
2. Li Guodong, Wang Wei, Li Fengjuan, et al. Study and Development of RO Membrane [J], Polymer Bulletin (China), 38 (2010)
3. Zhou Baiqing, Chen Zhihe. Water Treatment in Thermal Power Plant [M], China Electric Power Press, 177 (2009)
4. Wu Liancheng, Zheng Fei. Study of the Discharge and Utilization of RO Condensate [J], Henan Technology, **9**, 66 (2008)
5. Ning R Y, Tarquin A, TrzcinskiM C, *et al.* Recovery optimization of RO concentrate from desert wells [J], Desalination, 201, 315-322 (2006)
6. Zhang Sheng, Mao Yonghao, et al. Recycling of RO Condensate [J], Industrial Water & Wastewater, **38**, 56 (2007)
7. Song Sheling, Guo Lijun, Sun Xiufang. Recycling and Application of RO Wastewater [J], Energy Conservation, **6**, 39-40 (2005)
8. Zhang Deqiang, Qiao Baodong, Wei Bingju, et al. Study of Concentrated Seawater Salt Technology After Seawater Desalination [J], Journal of Salt and Chemical Industry, **6**, 4-9 (2007)
9. Yang Q F, Ma Z F, Hasson D, *et al.* Destruction of anti-scalants in RO concentrates by electrochemical oxidation [J], Journal of Chemical Industry and Engineering (China), **55**, 339-340 (2004)
10. Badruzzaman M, Oppenheimer J, Adham S, *et al.* Innovative beneficial reuse of reverse osmosis concentrate using bipolar membrane electrodialysis and electrochlorination process [J], Membr Sci, **326**, 392-399 (2009)
11. Wang Jun, Luan Zhaokun, *et al.* Study of Applying Hydrophobic Membrane Distillation Technology in RO Condensate Recycling [J], China Water & Wastewater, **23**, 1 (2007)
12. Drioli E, Macedonio F. New trends in membrane technology for water treatment and desalination[A]. 5th IWA Conference on Membranes for Water and Wastewater Treatment [C], Beijing: IWA-MTC 2009 Organising Committee (2009)
13. Ding Xuwei, Tan Qiu. Industrial Crystallization [M], Chemical Industry Press, **10**, 122-125 (1985)
14. Lu Jiufang. Separation Process Chemistry [M], Tsinghua University Press, 232 (1993)
15. Chen Ping. The study of water softening and fluorine removal through carrier-induced precipitation and crystallization [D], Xi'an University of Architecture and Technology, 7 (2004)

MATEC Web of Conferences **49**, 05001 (2016)
DOI: 10.1051/matecconf/20164905001

Fuels by Waste Plastics Using Activated Carbon, MCM-41, HZSM-5 and Their Mixture

Norbert Miskolczi [1], Chunfei Wu [2], Paul T. Williams [3]

[1] *Chemical Engineering and Process Engineering Institute, MOL Department of Hydrocarbon and Coal Processing, University of Pannonia, Veszprém, 10 Egyetem u., 8200, Hungary*
[2] *Energy Research Institute, University of Leeds, Leeds, LS2 9JT, UK*
[3] *School of Engineering, University of Hull, Hull, HU6 7RX, UK*

Abstract. Waste material was pyrolyzed in a horizontal tubular reactor at 530-540°C using different catalysts, such as activated carbon, MCM-41, HZSM-5 and their mixtures. Products were investigated by gas-chromatography, EDXRFS and standardized methods. Catalysts significantly affected the yields of volatiles; e.g. HZSM-5 catalyst increased especially the yield of gaseous hydrocarbons, while MCM-41 catalyst was responsible for increasing the pyrolysis oil yield. Synergistic effects were found using mixtures of different catalysts. Furthermore the catalysts modified the main carbon frame of the products. Pyrolysis oil obtained over HZSM-5 catalyst contained large amounts of aromatics, while MCM-41 catalyst mainly isomerized the carbon frame. Regarding contaminants it was concluded, that the sulphur content could be significantly decreased by activated carbon, however it had only a slight effect to the other properties of the products.

1. Introduction

Pyrolysis has benefits not only by reducing the huge mass of polymer wastes, but also enormous energy could be saved by chemical recycling. During pyrolysis, the polymer macromolecules decompose into gaseous and liquid products, which have lots of similarities with commercial fuel gas, naphtha or even gas oil. In general, 400-1000 °C is used for waste depolymerization depending on the technologies (using batch reactor, tubular reactor, fluidized bed, rotary kiln, etc.) [1-5]. Higher temperature promotes the formation of gaseous hydrocarbons, coke deposits, and the products contain lots of cyclic hydrocarbons or aromatics. Aliphatic hydrocarbons (both saturated and unsaturated) are the main products using lower pyrolysis temperatures (<600°C) [4-9]. The product yields and composition are affected also by the application of catalysts, which could increase the yield of volatiles (both gases and pyrolysis oils) [10-14]. It was demonstrated, that the structure of zeolites had a key role in relation to the product properties. E.g. mainly alkanes are produced over y-zeolite, while unsaturated hydrocarbons and aromatics are the main fractions using ZSM-5 [4, 10, 11, 16].

Based on references, the energetic application of pyrolysis product is the mostly investigated way for their further utilization [15, 17]. However there are different alternatives for further product utilization, depending on the pyrolysis parameters (Figure 1).

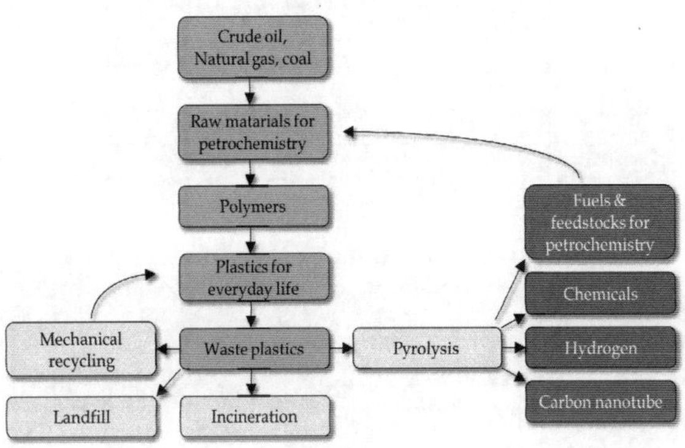

Figure 1. Ways for utilizations of pyrolysis products obtained by waste plastic

Not only fuels, but also different other valuable products could be generated: chemicals, hydrogen, olefin rich fractions for petrochemical application, carbon nanotube, carbon black, etc. Hydrogen rich gas is the main product, using metal containing catalyst and high temperature, while olefin rich fraction and other chemicals are the main products using catalyst with high Si/Al ratio. Carbon nanotube production also requires special catalyst and elevated temperature. It is important to remark that corrosion and emission problems could occur when contaminated pyrolysis products are further used. Therefore the decreasing of contaminants in products is a real problem [9].

In this work thermo-catalytic pyrolysis of waste low density polyethylene and high density polyethylene was investigated in a horizontal tubular reactor at 530-540°C using activated carbon, MCM-41 and HZSM-5 catalysts and their mixtures. Especially the effect of catalysts and their mixtures to the product yield and composition was investigated.

2. Material and methods

2.1 Raw materials

Mixture of real waste high density polyethylene (HDPE) and low density polyethylene (LDPE) from packaging was used in this work. According to preliminary analysis, raw material contained 74.8% HDPE and 15.2% LDPE. The main properties of the raw materials are summarized in Table 1.

Table 1. The main properties of real waste plastics

Property	Value
Average particle size, mm	3-4
Proximate analysis	
Ash content, %	2.9
Volatile content, %	1.4
Combustible, %	95.7
Ultimate analysis	
C, %	85.71
H, %	14.21
Cl, %	0.05
S, %	0.02

Samples of average particle size of 3-4mm were used. Owing to the real waste source, raw material had contamination, in the form of chlorine (0.05%) and sulfur (0.02%), while the moisture and ash content were 1.4% and 2.9%, respectively.

In order to modify the yield and composition of products, different catalysts and their mixtures were tested: activated carbon, MCM-41 and HZSM-5. Table 2 demonstrates the composition of catalyst mixtures.

Table 2. The composition of catalysts (in%)

Catalyst	Activated carbon	MCM-41	HZSM-5
C-1	100	-	-
C-2	-	100	-
C-3	-	-	100
C-4	50	50	-
C-5	50	-	50
C-6	-	50	50
C-7	33.33	33.33	33.33

The main properties of catalysts are summarized in Table 3. Catalysts had similar grain size, in the range between 10 and 109 m. MCM-41 had higher Si/Al ratio than HZSM-5, while its acidity was lower. Activated carbon and MCM-41 catalyst had the highest BET surface area with 859m^2/g and 824m^2/g, respectively.

Table 3. The properties of catalysts

	Activated carbon	MCM-41	HZSM-5
Grain size, m	10-109	15-84	10-79
Si/Al ratio	-	40	25
Acidity, mequiv of NH_3/g	-	0.15	0.60
BET area, m^2/g	859	824	298

2.2 Apparatus for waste plastic pyrolysis

A horizontal tubular reactor was used for waste plastic pyrolysis. Waste plastics were ground to particles of 3-4mm before the pyrolysis. A hopper was filled with the plastic in the first section of the electrically heated extruder. The feed rate was 2 kg hourly. The molten waste raw material was directly fed into the reactor stage from the extruder end section, where it was suddenly heated to 530-540°C. Waste HDPE and LDPE were decomposed into smaller hydrocarbon molecules in the reactor. The mixtures of decomposed hydrocarbons were separated into gases, pyrolysis oil and heavy oil in a distillation column connected to the end section of reactor. Gases from the top section of the distillation column were condensed in a water cooled tubular heat exchanger, and then a given ratio of the condensed product (pyrolysis oil) was driven back to the distillation unit top section. Higher efficiency of distillation could be reached by this procedure. The bottom temperature of the separation unit was set 300-350°C. Temperatures were controlled by PID controllers in all of the technology. In case of thermo-catalytic pyrolysis 4 wt% of catalysts were added into the raw material. Products were collected and further analyzed.

2.3 Methods

Gases and pyrolysis oil were analyzed by DANI gas-chromatograph using FID. For gas analysis 50 m x 0.32 mm fused silica column with Al_2O_3/KCl coating was used and the temperature was constant 40°C during the analysis. In case of pyrolysis oil analysis, 30m x 0.32mm Rtx®-1 column was used. The temperature was elevated from 40°C to 330°C with 15°C/min ramp, and then the temperature was kept constant for 10 minutes.

The concentrations of sulphur and chlorine were measured by a PHILLIPS PW 4025/02 (MiniPal) EDXRF spectrometer, which was fitted with a rhodium-side window tube anode (power of 9W) and Si-PIN detector.

Volatile characteristic of pyrolysis oil was also determined by Engler distillation test to measure the volume of hydrocarbon distilled at given temperatures (ASTM-D 1078).

3. Results and discussion

3.1 Product yields

Product yields are summarized in Figure 2, which shows significant differences among quantities of gases, pyrolysis oil and heavy oil in the presence of catalysts and their mixtures.

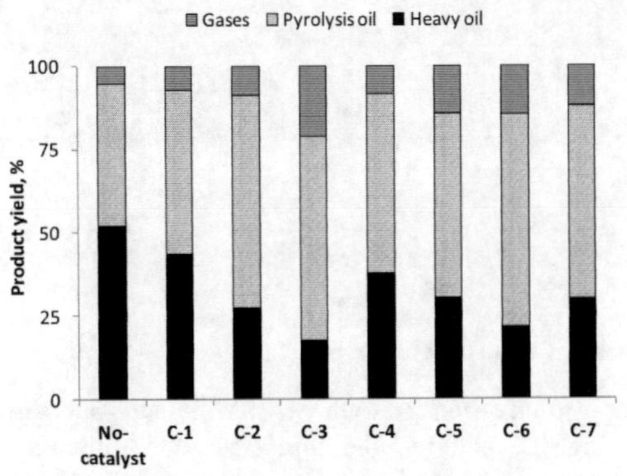

Figure 2. Product yields using different catalysts.

Pyrolysis of waste polymers resulted in 42.7% pyrolysis oil and 5.1% gases without catalyst, which did not change significantly using activated carbon, because the yields of gases and pyrolysis oil were only increased to 7.2% and 49.2%. Results demonstrated that the activated carbon had only a very slight catalytic effect. The increase in yields of volatiles rather could be attributed to the larger heat transmission area, which was due to the high BET surface area of activated carbon.

The increase in volatile yields was well shown using MCM-41 or HZSM-5 catalysts. HZSM-5 significantly increased the yields of gaseous product (21.1%), while 63.9% and 61.4% pyrolsis oil yield was observed using MCM-41 and HZSM-5 catalysts, respectively. Regarding the catalyst mixtures it could be concluded, that the adding of activated carbon only slightly deteriorated the advantageous properties of both MCM-41 and HZSM-5 to the pyrolysis oil yield increasing. Furthermore nearly the highest (63.7%) yield of pyrolysis oil was obtained from waste pyrolysis over 50%/50% MCM-41/HZSM-5 catalysts. MCM-41 catalyst had larger pore size than HZSM-5; therefore it has catalytic activity rather in pyrolysis oil increasing, while HZSM-5 catalyst with smaller pores efficiently could increase the yields of gaseous products.

3.2 Gases

The compositions of gaseous products are summarized in Tables 4/a and 4/b. Based on GC analysis, it is clear that the gaseous fraction contained methane, ethane and ethene, propene and propane, butane and butene, iso-C4, pentene and pentane, iso-C5 hydrocarbons. Without catalyst, the concentrations of before listed components were 13.5%, 29.9%, 22.8%, 28.6%, 3.4% and 0.9%, respectively. It means that mainly the C_2, C_3 and C_4 hydrocarbons represented the most dominant fractions in the gases. Similar result was found in case of catalysts; however notable isomerisation of the main carbon frame was also demonstrated. Especially the iso-C4 concentration could be increased; it was 11.7% and 16.6% using MCM-41 and HZSM-5 catalysts respectively. It is important to remark, that the application of activated carbon did not affect significantly the concentration of components in the gases.

Table 4/a. Composition of gases obtained by waste plastic pyrolysis at 530-540°C, in vol%.

Catalyst	**No-catalyst**	**C-1**	**C-2**	**C-3**
Methane	13.5	16.9	14.3	22.3
Ethene, Ethane	26.9	25.6	21.7	17.0
Propene, Propane	22.8	20.7	22.8	15.0
Butene, Butane	28.6	28.4	18.2	18.8
iso-C4	3.4	2.8	11.7	16.6
Pentene, Pentane	4.0	4.2	5.1	4.3
iso-C5	0.9	1.5	6.3	6.0

Regarding catalysts mixtures, it could be concluded, that only a slightly less effect of catalysts were found using 50% activated carbon and 50% MCM-41, or 50% activated carbon and 50% HZSM-5, than it was demonstrated in the case of MCM-41 and HZSM-5 catalysts. E.g. the iso-C4 content was 11.1% and 15.4% over 50% activated carbon and 50% MCM-41, or 50% activated carbon and 50% HZSM-5, respectively. Similar result was found regarding iso-C5 compounds, because it was increased from 0.9% (without catalyst) to 1.5% (activated carbon) – 6.3% (MCM-41). Regarding the gaseous fraction the HZSM-5 catalyst had a more significant isomerisation effect, than MCM-41 catalyst, which could be attributed to its BET area. The main frame of instable molecule fragments could be isomerized in the propagation reaction step resulting in high concentration of branched molecules.

Table 4/b. Composition of gases obtained by waste plastic pyrolysis at 530-540°C, in vol%. (cont.)

Catalyst	**C-4**	**C-5**	**C-6**	**C-7**
Methane	13.6	19.8	16.1	13.7
Ethene, Ethane	20.8	21.0	19.6	23.1
Propene, Propane	19.6	16.4	20.6	21.3
Butene, Butane	25.6	21.0	18.8	17.0
iso-C4	11.1	15.4	13.8	13.0
Pentene, Pentane	3.8	3.7	4.9	6.3
iso-C5	5.5	2.7	6.2	5.6

3.3 Pyrolysis oil

Pyrolysis oils were analyzed by gas-chromatography to know more about their composition. Based on the results it was concluded, that C_5-C_{34} hydrocarbons were represented in the pyrolysis oils. The composition of pyrolysis oils as function of carbon atoms are shown in Figure 3. Owing to the structure of HDPE and LDPE, two well separated peaks were appeared in gas chromatograms belong to each carbon atoms. The statistical distribution of compounds concentration as function of carbon atoms is also characteristic of polyethylene degradation. The structure of polyethylene is relative stable, because it contains mainly secondary carbon atoms, therefore the distortion of the electron cloud and the parts with structural faults are relatively low.

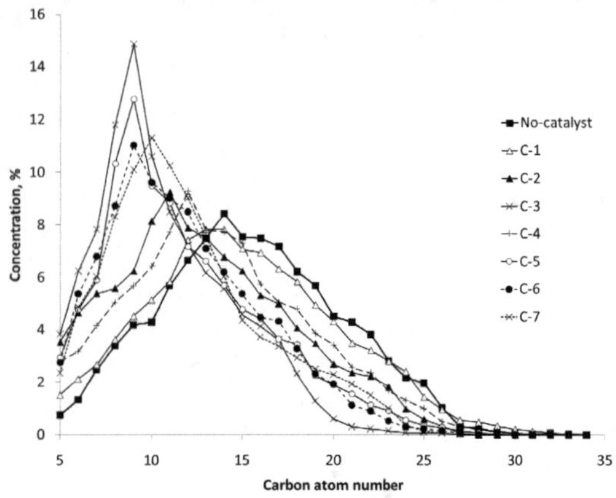

Figure 3. Distribution of compounds in pyrolysis oil as function of carbon atoms.

Regarding thermal pyrolysis without catalyst, compounds with carbon atoms of 14 had the highest concentration (8.43%). The application of activated carbon did not change notably the distribution curve, because e.g. the maximum point also belongs to carbon atoms of 14 (7.82%). However both MCM-41 and HZSM-5 catalysts had a significant effect on the product distribution: e.g. the maximum points were shifted to carbon atoms of 11 (9.02%) and 9 (14.88%) using MCM-41 and HZSM-5 catalysts, respectively. Regarding HZSM-5 catalyst, the high aromatic portion of pyrolysis oil was the cause for the high value of C_9 compounds.

Significant increases in compounds with shorter chain ($<C_{14}$) was found using both MCM-41 and HZSM-5 catalysts, while the concentration of larger molecules ($>C_{14}$) decreased in those cases. That was the cause for the average molecular weight decreasing using both MCM-41 and HZSM-5 catalysts (Figure 4). As the data show the average molecular weight of pyrolysis oil was 242g/mol without catalyst, while it could be significantly decreased by both MCM-41 (178g/mol) and HZSM-5 (139g/mol) catalysts, while it could be only slightly decreased by activated carbon (235g/mol). Regarding the mixtures of catalysts, the 50%/50% HZSM-5 and activated carbon mixture have shown also a significant effect resulting in molecular weight decrease (142g/mol).

Figure 5 shows the composition of pyrolysis oils, which contain non-branched, branched alkanes and aromatic hydrocarbons. Results well demonstrate the isomerisation and aromatization effect of catalysts. E.g. pyrolysis oil contained 89.2% non-branched and 10.8% branched hydrocarbons without catalyst. The concentration of branched hydrocarbon increased to 31.9% and 35.5% by the using of MCM-41 and HZSM-5 catalysts.

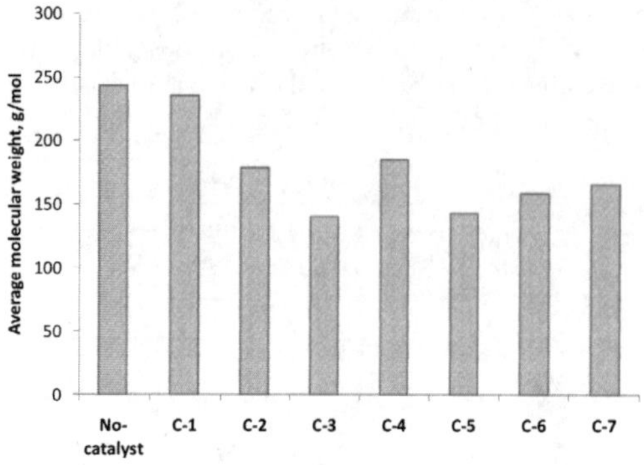

Figure 4. Calculated average molecular weights of pyrolysis oil.

Regarding aromatics, the HZSM-5 catalyst showed the highest activity, because the concentration of aromatic hydrocarbons reached 34.1% using HZSM-5. This result was confirmed by earlier work, when the high activity of HZSM-5 catalyst to the aromatization reactions was concluded [4,10,11,16]. It is also clear, that activated carbon could not increase the aromatic content in pyrolysis oil. MCM-41 catalyst rather had an effect to the isomerisation reactions, which resulted in an elevated concentration of branched compounds. Regarding the application of catalyst mixtures, aromatization and isomerisation effect was found in the presence of HZSM-5 catalyst.

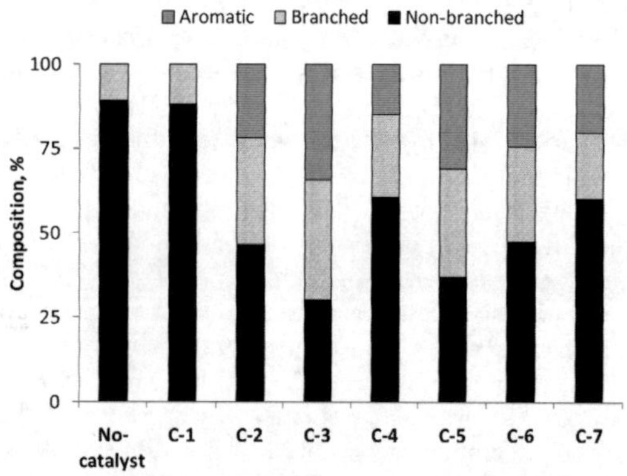

Figure 5. Composition of pyrolysis oils.

The composition of aromatics in pyrolysis oils are summarized in Figure 6. Pyrolysis oils contain benzene, toluene, ethyl-benzene, styrene, xylenes, isopropyl-benzene, a-methyl styrene, cumene and other aromatic hydrocarbons. According to the results, over HZSM-5 catalyst (C-3) significant values of ethyl-benzene, styrene and xylenes were measured; 17.12%, 19.67% and 16.33%, respectively. MCM-41 catalyst (C-2) has also aromatization function; because the total aromatic concentration was 21.6% using that catalyst.

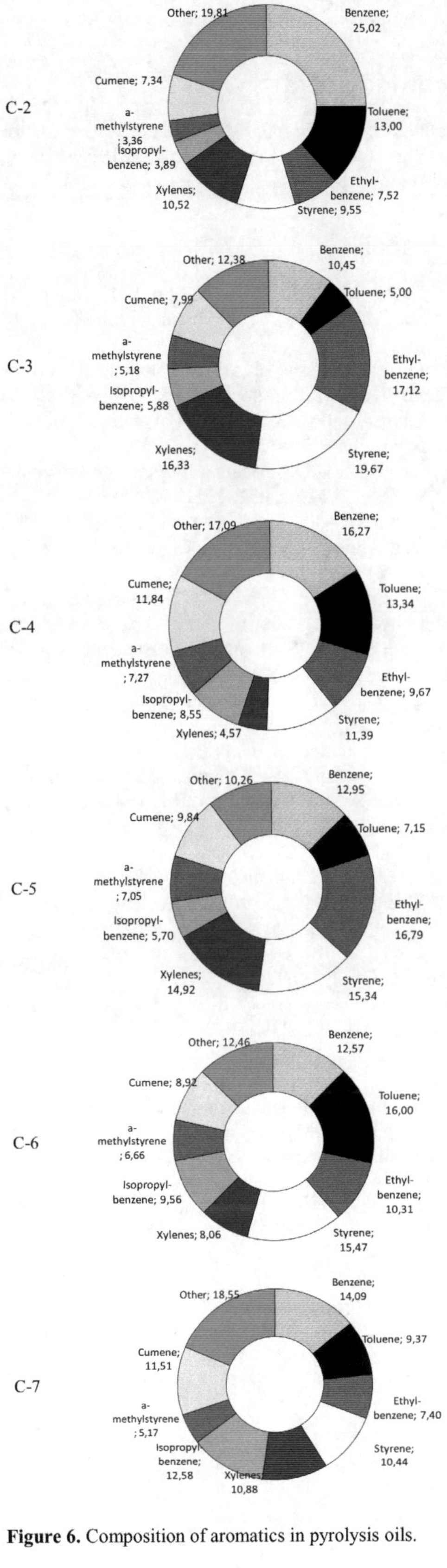

Figure 6. Composition of aromatics in pyrolysis oils.

It is an important observation, that the MCM-41 catalyst increases the concentration of benzene and toluene (25.05% and 13.00%), while that of ethyl-benzene, styrene and xylenes were lower (7.52%, 9.55% and 10.52%). This phenomenon could be attributed to the larger pore size and BET surface area of the MCM-41 catalyst, because those properties are favoured for the hydrogen transfer reactions and aromatization. That is why the aromatic side chain of aromatics could be further cracked which resulted in smaller molecules with shorter side chain. Furthermore also further decomposition of side chain if aromatics were found in the case of catalyst mixtures using MCM-41 catalyst (C-4, C-6).

As it was demonstrated in Table 1, the raw material contained 0.05% chlorine and 0.02% sulphur, which should be transformed into compounds, which appeared in the pyrolysis oil fractions. Figure 7 demonstrates the concentrations of chlorine and sulphur in the pyrolysis oils. Without catalysts 257ppm chlorine and 125ppm sulphur were measured, which could be decreased using catalysts. Regarding chlorine, the MCM-41 catalyst had the highest activity in decreasing, because the Cl concentration was122ppm. Activated carbon had shown only a slight effect to the volatile product increasing and modifying the pyrolysis oil composition; however it has a significant effect in the case of sulphur content. Pyrolysis oil obtained by the degradation over activated carbon contained only 31ppm sulphur, which is a dominant decrease. On the other hand the two other catalysts only slightly could decrease the S content; 79ppm and 81ppm using MCM-41 and HZSM-5 catalysts. Presumably the large BET surface area was the cause for both Cl and S decreasing in the case of MCM-41 and activated carbon. According to Table 1, activated carbon had 859m^2/g, while MCM-41 catalyst 824m^2/g BET surface area. Therefore possibly physical sorption was the cause for both Cl and S decreasing.

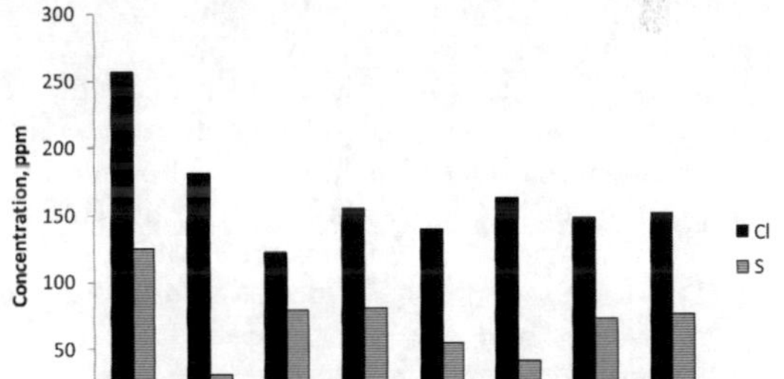

Figure 7. Chlorine and sulphur content of pyrolysis oils.

Volatile characteristic of pyrolysis oils was also followed by Engler distillation test. Figure 8 summarizes the distillation curves. The data well demonstrates, that pyrolysis oil obtained by catalyst free experiment contain nearly half of diesel oil range and 25%-25% kerosene and naphtha range hydrocarbons. In the presence of catalysts (especially HZSM-5) the portions of naphtha-like and kerosene-like hydrocarbons were higher, e.g. using HZSM-5 they were nearly 60% and 25%, respectively. It was confirmed by earlier work, because it was demonstrated that owing to the small pore size, the HZSM-5 catalyst results especially hydrocarbons in naphtha boiling point range (Bp below 180°C) [10,11,16].

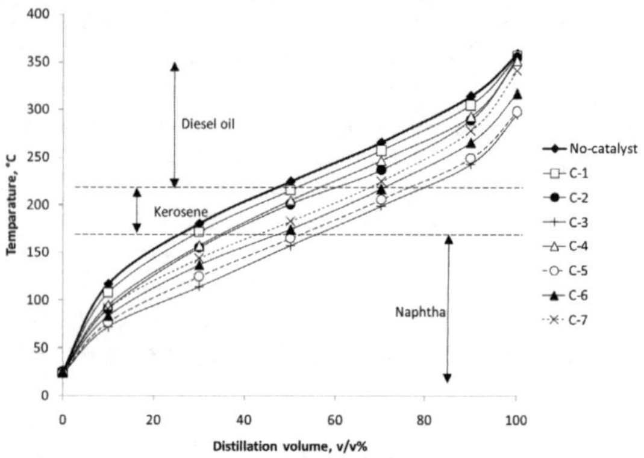

Figure 8. Engler distillation curves of pyrolysis oils.

Activated carbon had only a slight effect on the boiling point range, because the temperatures belong to specific distillation volumes, was only slightly lower than that without catalyst. MCM-41 catalyst could also slightly affect the boiling points, but its effect was most notable than in the case of activated carbon. Regarding catalyst mixtures, it was concluded, that especially the HZSM-5 containing mixtures (C-5, C-6 and C-7) could significantly increase the yield of naphtha or kerosene and decrease the yield of diesel oil.

4 Conclusion

The yields of gaseous product could be increased especially by HZSM-5 catalyst, while the yield of pyrolysis oil was increased both by HZSM-5 and MCM-41 catalyts. Activated carbon had only a slight effect on the yields and hydrocarbons in the product. Significant isomerization effect was demonstrated both in the gases and pyrolysis oils using HZSM-5 catalyst and HZSM-5 catalyst containing catalyst mixtures. MCM-41 and HZSM-5 catalyst had also an aromatization effect, and the MCM-41 catalyst with larger pore size had activity in the formation of benzene and toluene. The most benefit of activated carbon was the enormous decrease in sulphur content, while the chlorine could be decreased especially by MCM-41 catalyst, which was attributed to its large pore surface area. Volatile characteristic of pyrolysis oils demonstrated the increase of the naphtha-like fraction content of pyrolysis oil using HZSM-5 catalyst.

Acknowledgment

The authors acknowledge the Horizon 2020, Marie Curie Research and Innovation Staff Exchange (RISE) (MSCA-RISE-2014 (Flexi-pyrocat, No.: 643322)).

References

1. H. T. Lin, M. S. Huang, J. W. Luo, L. H. Lin, C. M. Lee, and K. L. Ou, Fuel Processing Technology, **91**, 1355 (2010)
2. E. V. Antonakou, K. G. Kalogiannis, S. D. Stefanidis, S. A. Karakoulia, K. S. Triantafyllidis, A. A. Lappas, and D. S. Achilias, Polymer Degradation and Stability, **110**, 482 (2014)
3. Y. H. Lin, M. H. Yang, Journal of Molecular Catalysis A: Chemical, **231**, 113 (2005)
4. G. Elordi, M. Olazar, M. Artetxe, P. Castano, and J. Bilbao, Applied Catalysis A: General, 415-416, 89-95 (2012)
5. Y. H. Lin, H. Y. Yen, Polymer Degradation and Stability, **89**, 101 (2005)
6. W. C. Huang, M. S. Huang, C. F. Huang, C. C Chen, K. L. Ou, Fuel, **89**, 2305 (2010)
7. S. S. Kima, S. Kim, Chemical Engineering Journal, **98**, 53 (2004)
8. N. S. Akpanudoh, K. Gobin, G. Manos, Journal of Molecular Catalysis A: Chemical, **235**, 67 (2005)
9. X. Yang, L. Sun, J. Xiang, S. Hu, and S. Su, *Waste Management* 33 (2013), p. 462
10. J. F. Mastral, C. Berrueco, M. Gea, J. Ceamanos, Polymer Degradation and Stability, **91**, 3330 (2006)
11. X. Zhang, H. Lei, G. Yadavalli, L. Zhu, Y. Wei, Y. Liu, Fuel, **144**, 33 (2015)
12. J. Bozi, M. Blazsó, Journal of Analytical and Applied Pyrolysis, **97**, 189 (2012)
13. W. L. Fanchiang, Y. C. Lin, Applied Catalysis A: General, 419-420, 419 (2012)
14. R. A. García, D. P. Serrano, D. Otero, Journal of Analytical and Applied Pyrolysis, **74**, 379 (2005)
15. M. R. Jan, J. Shah, H. Gulab, Fuel, **105**, 595 (2013)
16. A. López, I. Marco, B. M. Caballero, M. F. Laresgoiti, A. Adrados, and A. Aranzabal, Applied Catalysis B: Environmental, **104**, 211 (2011)
17. K. H. Lee, Polymer Degradation and Stability, **93**, 1284 (2008)

MATEC Web of Conferences **49**, 05002 (2016)
DOI: 10.1051/matecconf/20164905002

Thermo-Catalytic Pyrolysis of Waste Plastics from End of Life Vehicle

Norbert Miskolczi [1, a], Zsuzsanna Czégény [2]

[1] *Chemical Engineering and Process Engineering Institute, MOL Department of Hydrocarbon and Coal Processing, University of Pannonia, Veszprém, 10 Egyetem u., 8200, Hungary*
[2] *Institute of Materials and Environmental Chemistry, Research Centre for Natural Sciences, Hungarian Academy of Sciences, Budapest 1117 Magyar Tudósok krt. 2, Hungary*

Abstract. Pyrolysis of waste plastics is widely used recycling method. Owing to the end-of-life vehicles regulations, 95% of passenger cars and vehicles must reused/recovered after the dismantling. Pyrolysis of waste polyethylene and polypropylene obtained from end-of-life vehicles was investigated in a continuously stirred batch reactor using 500 and 600°C temperatures. To ensure the pyrolysis reactions the tested catalysts (5% of ZSM-5, HZSM-5, Ni-ZSM-5 and Fe-ZSM-5) were added directly to the mixtures of raw materials. Products of pyrolysis were separated into gases, pyrolysis oil and heavy oil, which was further analyzed by gas-chromatography, Fourier transformed infrared spectroscopy and other standardized methods. Based on the results it was concluded, that the catalysts significantly increase the yields of volatile products, and modify their composition. Especially the alkane/alkene ratio, the methane concentration and the concentration of branched hydrocarbon could be affected by the applied catalysts. Ni-ZSM-5 catalyst had the highest activity in methane production, while HZSM-5 catalyst proved effective in isomerization reactions. Using H-ZSM-5, Ni-ZSM-5, and Fe-ZSM-5 catalyst notably decreased average molecular weight of pyrolysis oils and significantly higher aromatic content was observed.

1 Introduction

Waste polymers could be transformed into valuable hydrocarbons by pyrolysis. The effect of the process parameters to the product properties is well published [1-5]. E.g. higher gas and oil yields are obtained using higher temperature and catalyst. Another important parameter is the kind of raw material, because it has a dominant effect on the product yield and compositions. Based on references, especially the decomposition of virgin polymers was investigated and less information is available related to the pyrolysis of real wastes.

Modern cars and vehicles contain approximately 10-15% plastics, depending on their type. Regarding the type of plastics, both mass plastics and engineering plastics are widespread used in automotive sector. Especially polyethylene, polypropylene, polyamide, polyurethane, acrylonitrile butadiene styrene, etc. are used. Figure 1 demonstrates the average composition of parts of vehicles [6-8]. In general, the further utilization of metal and glass content of the vehicles are solved problem, however the long time sustainable plastic utilization is an open question. Regarding waste plastics from ELV, polyethylene, polypropylene and polystyrene are the mostly favored for pyrolysis, because the chemical structure build up only from carbon and hydrogen; therefore their pyrolysis oil also free of unwanted components.

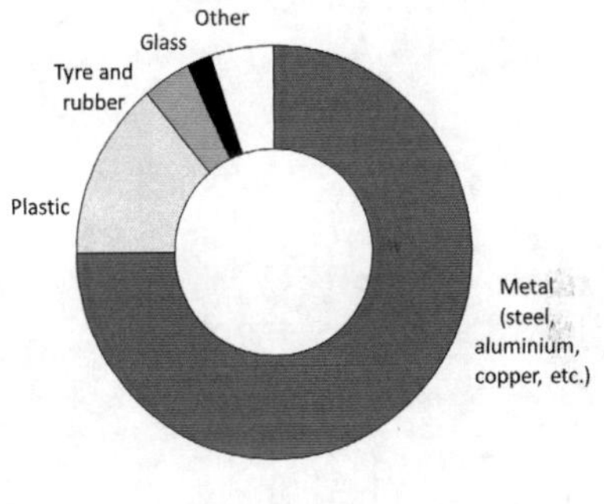

Figure 1. Average composition of vehicle parts.

Based on references, especially the pyrolysis of waste oils and tires from automotive sector are investigated. It was concluded, that both tires and waste oils should be advantageously used for hydrocarbon or energy production [9-11]. However aromatics, polyaromatics or even contaminants, such as sulfur in case of waste tire pyrolysis can cause serious problems during further utilization. Results of the investigation of the pyrolysis of waste plastics from automotive industry using pyrolysis–gas chromatography/mass spectrometry were published recently, which looks an attractive method for investigation of small amount of samples [12-14].

In our current work larger amount of waste polyethylene and polypropylene mixture from ELV was pyrolyzed using different catalysts (ZSM-5, HZSM-5, Ni-

[a] Corresponding author: mnorbert@almos.uni-pannon.hu

ZSM-5 and Fe-ZSM-5) in a continuously stirred batch reactor using 500 and 600°C temperature.

2 Material and methods

2.1 Raw materials

Waste high density polyethylene (HDPE), low density polyethylene (LDPE) and polypropylene (PP) were used as raw materials in our current work from ELV dismantling. ELV wastes contained 4.85% ash, especially glass fiber and 2.17% volatiles, therefore the combustibles were 92.98%.

Different catalysts were used for modifying the pyrolysis characteristic and properties of the products: ZSM-5, HZSM-5, Ni-ZSM-5 and Fe-ZSM-5. Wet impregnation was used for Ni-ZSM-5 and Fe-ZSM-5 synthesis, using $Ni(NO_3)_2 \cdot 6H_2O$ and $Fe(NO_3)_3 \cdot 9H_2O$, respectively, while 1M NH_4NO_3 was used for HZSM-5 synthesis at 80°C. Catalysts were dried for 10 hours at 110°C, and then each of them was calcined at 600°C for 3 hours.

2.2 Apparatus for waste plastic pyrolysis

A stainless steel continuously stirred batch reactor was used for pyrolysis of waste plastics from ELV using temperatures of 500 and 600°C. The schematic layout of the pyrolysis rig is shown in Figure 2.

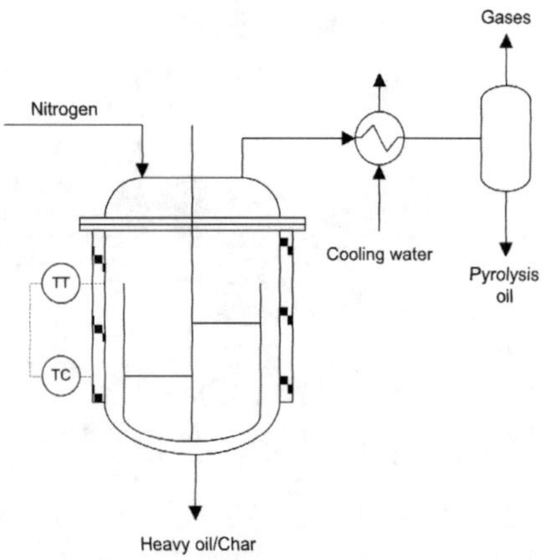

Figure 2. Batch apparatus for ELV waste plastic pyrolysis.

300g of raw materials and 5% of catalysts were weighted into the reactor. Then the reactor top section was placed and valve for nitrogen flush was opened. The temperature increased to the set value in 15 minutes. The temperature was measured on the reactor wall and controlled by a computer programmed controller. Mixtures of the decomposed hydrocarbons went through a condenser and then it was separated into gaseous and pyrolysis oil for further analysis.

2.3 Methods

Gases were analyzed by DANI GC using 50 m x 0.32 mm fused silica column with Al_2O_3/KCl coating.

Pyrolysis oils were also analyzed by gas-chromatography using a 30m x 0.32mm Rtx®-1 (nonpolar phase; Crossbond® dimethyl polysiloxane) column.

The saturated/unsaturated ratio of the oils was followed by a TENSOR 27 type FTIR spectrometer (resolution: 2 cm^{-1}, illumination: SiC Globar light, monocromator: Littrow prism, detector: RT-DLaTGS (Deutero L-α-Alanine Triglycin-Sulphate) type detector with KBr window) in the 4000-400 cm^{-1} wave number ranges.

Volatile characteristic of pyrolysis oil was determined by Engler distillation test (ASTM-D 1078), while the density was measured according to MSZ EN ISO 12185 standard.

Heavy oil fractions were analyzed by the following methods: flash point determination (ISO 2719:2002 and MSZ 15967:1979), determination viscosity (MSZ ISO 3105:1998), melting point determination (MSZ 3252:1973), density measurement (MSZ EN 1285:1996), and heating value (MSZ 24000-5:1978).

3 Results and discussion

3.1 Product yields

Gases, pyrolysis oil and heavy oil (together with solid char) had been obtained from pyrolysis experiments; product yields are shown in Figure 3/a and b.

Based on results, it is clear, that higher yields of gases and pyrolysis oil were obtained at higher temperature using 15 min reaction time, than that of at lower temperature after 25min pyrolysis, which was attributed to the higher thermal stability of C-C bonds at lower temperature. Catalysts also had a significant effect to the yields of product. Each of them could increase both the gaseous and pyrolysis oil yields.

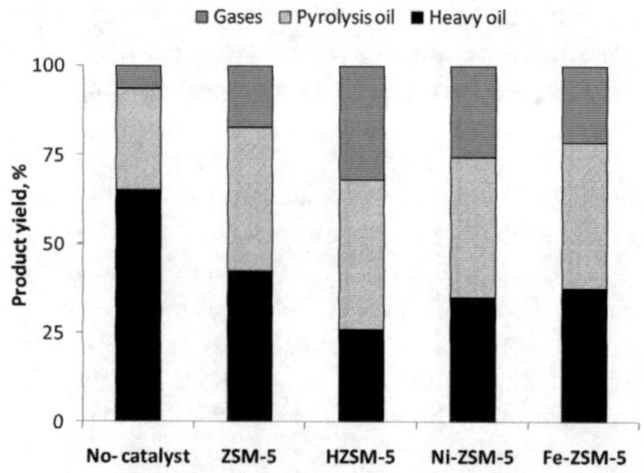

Figure 3/a. Product yields obtained at 500°C after 25min reaction time.

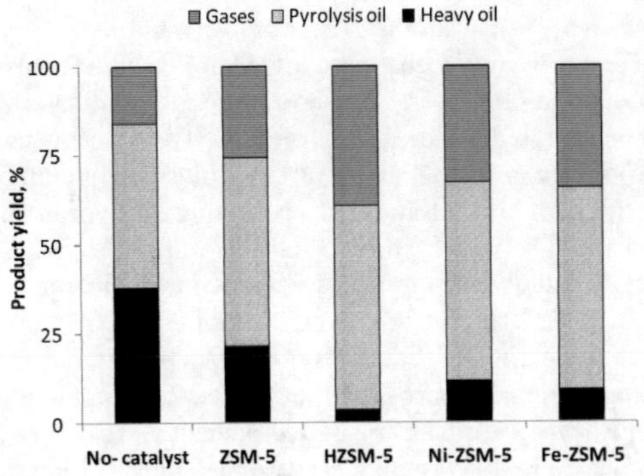

Figure 3/b. Product yields at 600°C after 15min reaction time.

The summarized yields of gases and pyrolysis oil was 34.7% and 62.2% at 500°(after 25min.) and 600°C (after 15min.), respectively. HZSM-5 catalyst showed the highest activity in yield increasing of gases and pyrolyis oil. The yield of pyrolysis oil was 40.3% and 17.2% gases were found at 500°C or 52.7% pyrolysis oil and 25.7% gases at 600°C using ZSM-5 catalyst. Both could be further increase using HZSM-5 catalyst, from 17.2% to 31.8% at 500°C and from 15.9% to 25.7% at 600°C in case of gases, while only 2-5% increasing was measured regarding pyrolysis oil using HZSM-5. It means that the protonated form of ZSM-5 catalyst had activity especially in increasing of gaseous products, which was the consequence of the relatively small pore size and large microporous surface are of ZSM-5 zeolite structure. Ni-ZSM-5 and Fe-ZSM-5 had similar affect on the yields of gases and pyrolysis oil as ZSM-5 catalyst, but the yield of gaseous products was higher (25.6% and 21.3% at 500°C and 32.5% and 34.2% at 600°C). Results also well show, that the volatile yield was higher at lower temperature, than that of using higher pyrolysis temperature, in case of each investigated catalyst. It refers that the radical initiated thermal cracking was more dominant at higher temperature, than the catalytic cracking with ionic mechanics.

3.2 Gases

Compositions of gases are shown in Figure 4/a and b. Gases contained aliphatic olefins, aliphatic paraffins, branched olefins, and branched hydrocarbons with carbon atoms up to six.

Without catalysts the amount of n-alkanes and n-alkenes are 43.5% and 31.1% at 500°C, while these values are 29.5% and 41.9% at 600°C, respectively. It means that higher n-alkane/n-alkene ratio was found at lower temperature, which was the consequence of more dominant reactions at 600°C. -scission is a well known decomposition mechanism of saturated hydrocarbon chain polymers, leading the formation of vinyl group terminated product. This decomposition route seems more significant at higher temperature. ZSM-5 catalyst increased especially the alkenes content, both branched and non-branched. E.g. the ratio of alkane/alkene was 1.3 and 0.78 at 500°C and 600°C, respectively; which could be decreased to 0.55 and 0.75, respectively. It is clear that the effect of the catalyst was more significant at lower temperature. The isomerisation of main carbon frame was moderate using ZSM-5. However, significant isomerisation effect was concluded using HZSM-5, because the non-branched/branched ratio could be changed from 4.34 to 1.06 at 500°C, while from 4.11 to 0.79 at 600°C temperature. Therefore the isomerisation effect of HZSM-5 catalyst was more significant at higher temperature.

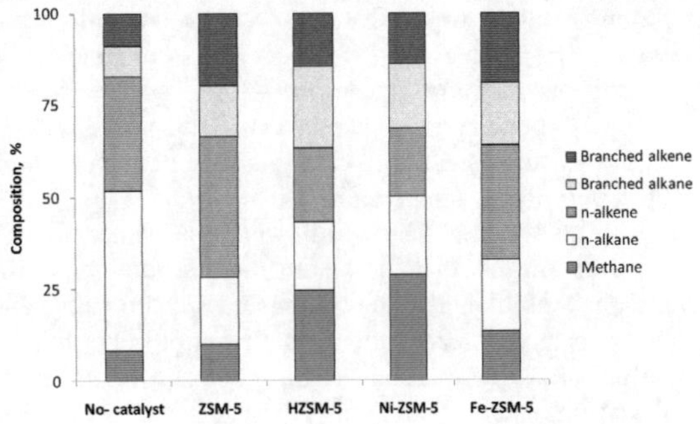

Figure 4/a. Composition of gases at 500°C after 25min reaction time.

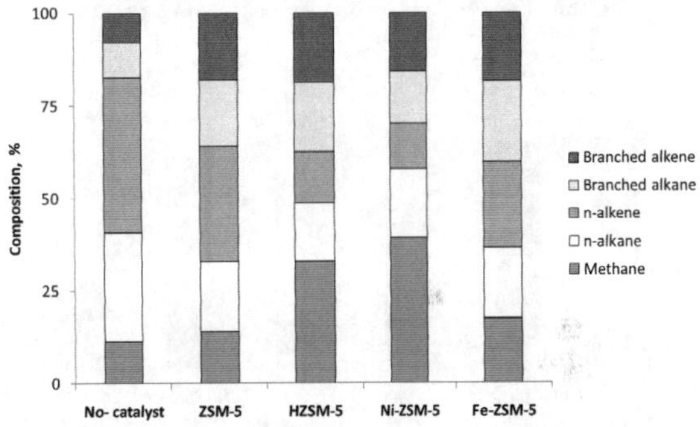

Figure 4/b. Composition of gases at 600°C after 15min reaction time.

Regarding HZSM-5 and Ni-ZSM-5 catalyst, increasing methane concentration was observed at higher temperature. Fe-ZSM-5 catalyst showed also an activity in olefin production reactions, but only at 500°C, while at 600°C the alkane/alkene ratio did not changed significantly comparing to the ratio obtained without catalyst. The methane concentration was 8.2% and 11.2% without catalyst at 500°C and 600°C, respectively. Those values could be increased to 28.7% and 34.7% by the using of Ni-ZSM-5 catalyst or 24.5% and 32.9% using HZSM-5 catalyst at 500°C and 600°C, respectively.

3.3 Pyrolysis oil

The composition of pyrolysis oil was investigated by gas-chromatography. The yield of n-alkane, n-alkene, branched alkane, branched alkene, cyclic and aromatic components were summarized, and presented in Figures 5/a and b. The pyrolysis oil contained only n-alkane, n-alkene, branched alkane and branched alkene when was pyrolysed without catalyst at lower temperature, while 14.2% aromatics and cyclic compounds was measured in the oil fraction obtained at 600°C. According to the literature, the cyclization and aromatization reactions became favourable at higher pyrolysis temperatures. In the presence of all investigated catalysts the yields of aromatics, cyclic and branched compounds significantly increased. It is interesting result, that the isomerisation effect of the catalysts was lower at higher temperature. The concentration of branched hydrocarbons obtained without catalyst was 27.9% and 26.4% at 500°C and 600°C, respectively. Pyrolysis oils obtained by experiments at 500°C contained 41.7%, 57.2%, 52.2% and 48.1% branched compounds using ZSM-5, HZSM-5, Ni-ZSM-5 and Fe-ZSM-5 catalysts, respectively. Using 600°C pyrolysis temperatures, that value was 33.1%, 42.5%, 40.3% and 42.9% following the same catalyst order. It means that the isomerisation efficiency of catalysts was higher at lower temperature. Probably due to the catalyst deactivation, the reaction rate of the thermal cracking was higher, than the rate of the ionic mechanism catalytic reactions at higher temperature. The coke formation is increasing with pyrolysis temperature, and the coke should be deposited on the catalyst surface, decreasing the catalyst activity.

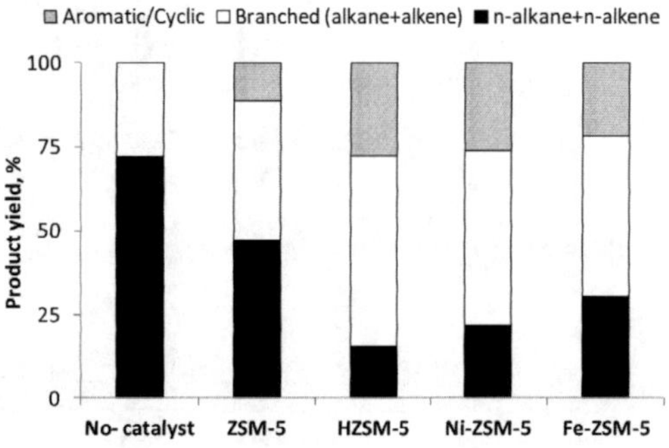

Figure 5/a. Composition of pyrolysis at 500°C after 25min reaction time.

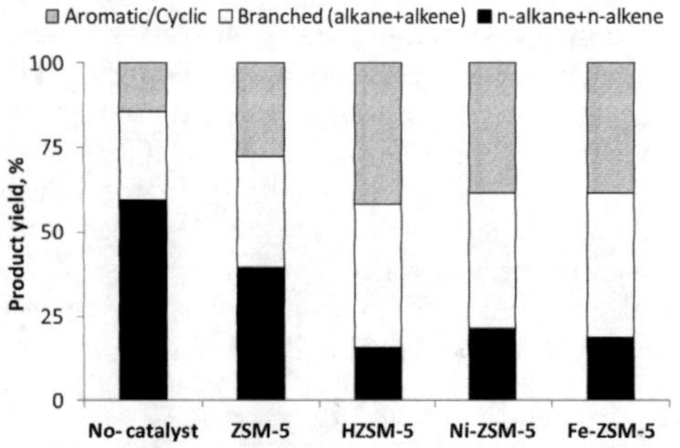

Figure 5/b. Composition of pyrolysis oils at 600°C after 15min reaction time.

Regarding aromatization it could be concluded, that the concentration of aromatics and cyclic compounds was nearly double at 600°C, than that of obtained at 500°C. The highest aromatic content was measured over HZSM-5 and Ni-ZSM-5 catalysts: 27.3% and 26.1% at 500°C or 41.6% and 39.2% at 600°C.

Figure 6 shows the ratio of alkanes and alkenes at 500°C and 600°C. As data shows, the ratios of saturated and unsaturated hydrocarbons were 1.64 and 1.19 without catalysts using 500°C and 600°C pyrolysis temperatures. The ratio of the unsaturated and saturated hydrocarbon yield could be increased by catalysts; especially Ni-ZSM-5 catalyst showed high activity in olefin production with ratios of 0.32 (at 500°C) and 0.31 (at 600°C). Transition metal containing catalysts has high activity in hydrogenation/dehydrogenation reactions, therefore alkenes were formed by dehydrogenation reactions of saturated hydrocarbon. Regarding the effect of temperature, results also well show that lower ratio was found at 600°C, which was caused by the higher formation of olefins by -scission reactions. Especially the catalyst free pyrolysis have resulted significantly lower value of ratio at higher temperature, because it was 1.94 at 500°C and 1.19 at 600°C. This phenomenon was attributed to the cracking of more C-C bonds at higher temperature.

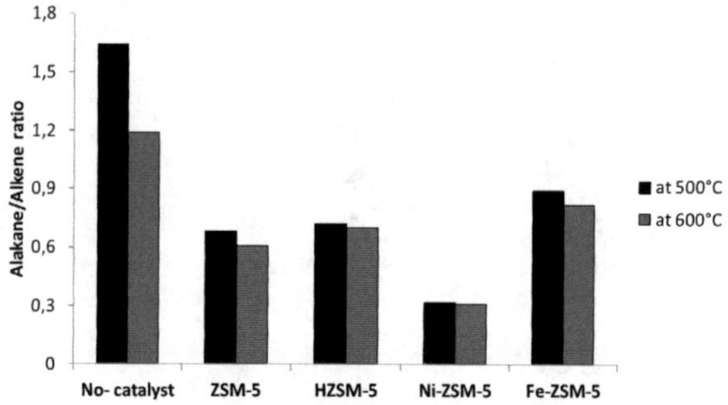

Figure 6. Alkane/alkene ratio.

Other properties of pyrolysis oils are summarized in Table 1, such as density, average molecular weight and boiling point ranges. Based on volatile test, pyrolysis oil contained hydrocarbons in gasoline and middle distillate boiling point ranges (22-331°C).

Table 1/a. Other properties of pyrolysis oils obtained at 500°C

Property	No-catalyst	ZSM-5	H-ZSM-5	Ni-ZSM-5	Fe-ZSM-5
Density at 20°C, g/cm^3	0.802	0.741	0.728	0.746	0.748
Average molecular weight, g/mol	221	137	116	142	149
Boiling points, °C					
IBP	25	25	24	26	22
50%	209	157	142	166	168
FBP	326	286	287	285	288

The density of pyrolysis oils was 0.802g/cm^3 and 0.788g/cm^3 at 500 and 600°C. It is clear that more C-C bonds of the main carbon chain should be cracked at higher temperature; therefore the density of pyrolysis oil was lower at higher temperature. Catalysts also decreased the average molecular weight of the pyrolysis oils, especially using HZSM-5 catalyst. E.g. without catalyst the average molecular weight was 221g/mol, while using HZSM-5 at 500°C it decreased to 116g/mol. On the other hand both density and molecular weight decreased at 600°C, comparing to 500°C pyrolysis temperature. Similar trend was observed regarding the boiling point ranges, because it was found that larger distillation volume belongs to same boiling points using catalysts than without them.

Table 1/b. Other properties of pyrolysis oils obtained at 600°C

Property	**No-catalyst**	**ZSM-5**	**H-ZSM-5**	**Ni-ZSM-5**	**Fe-ZSM-5**
Density at 20°C, g/cm^3	0.788	0.735	0.722	0.730	0.738
Average molecular weight, g/mol	203	124	102	120	126
Boiling points, °C					
IBP	25	27	22	24	25
50%	199	151	137	149	160
FBP	331	275	273	278	279

3.4 Heavy oil/Char

Heavy oil was the bottom product of pyrolysis obtained from the pyrolyis reactor at the end of reaction. This fraction was viscous light or dark brown products. In general they were solid at room temperature, because the melting point was in the range of 39 and 56°C. Table 2/a and 2/b summarizes the main properties of heavy oil obtained by pyrolysis at 500 and 600°C. Because of more C-C scission, heavy oil have lower density, melting point, and flash point obtained at 600°C, than 500°C, furthermore the listed properties were also lower using catalysts.

Heating values changed between 40.5 and 41.1MJ/kg at 500°C or between 41.0 and 41.4MJ/kg at 600°C. It means, that heavy oil had enough high heating values for their further fuel-like utilization, especially for heat, steam or even electricity production. Pyrolysis in the presence of HZSM-5 catalyst resulted the lowest lower density, melting point, and flash point, which was attributed to the cracking reactions on the large surface of HZSM-5 and its high acidity at 500°C. Experiments over HZSM-5 catalyst at 600°C did not result heavy oils product, only char, because the catalyst was too active and fully cracked the polymer main chain.

Table 2/a. Main properties of pyrolysis oils obtained at 500°C

Property	**No-catalyst**	**ZSM-5**	**H-ZSM-5**	**Ni-ZSM-5**	**Fe-ZSM-5**
Apparence	Dark brown solid	Light brown solid	Light brown solid	Light brown solid	Light brown solid
Density*, g/cm^3	0.847	0.826	0.823	0.830	0.839
Melting point, °C	56	51	46	51	53
Flash point, °C	208	184	175	179	197
Heating value, MJ/kg	40.9	41.1	40.5	40.7	41.0
Viscosity**, mm^2/s	5.2	4.5	4.2	4.3	4.4

* at 20°C, ** at 70°C

Table 2/b. Main properties of pyrolysis oils obtained at 600°C

Property	**No-catalyst**	**ZSM-5**	**Ni-ZSM-5**	**Fe-ZSM-5**
Apparence	Light brown solid	Light brown solid	Light brown solid	Light brown solid
Density*, g/cm^3	0.831	0.801	0.914	0.809
Melting point, °C	45	39	39	40
Flash point, °C	195	189	187	190
Heating value, MJ/kg	41.0	41.1	41.4	41.2
Viscosity**, mm^2/s	4.6	3.8	3.6	3.7

* at 20°C, ** at 70°C

4 Conclusion

Waste plastics from ELV dismantling have been pyrolyzed in a continuously stirred batch reactor at 500 and 600°C using different catalysts.

- Significant increasing in gases and pyrolysis oils was found using modified ZSM-5 catalysts. HZSM-5 catalyst had high activity in formation of gaseous product.
- High concentration of methane was found using HZSM-5 and Ni-ZSM-5 catalysts.
- Mainly branched hydrocarbons were obtained over HZSM-5 catalyst, which was the consequence of the isomerization effect.
- ZSM-5 and Fe-ZSM-5 catalyst had high activity in formation of unsaturated hydrocarbons.
- The pyrolysis oil contained n-alkane, n-alkene, branched alkane, branched alkene, cyclic and aromatic hydrocarbons, and the composition was affected by

catalysts, especially the aromatic and branched hydrocarbon content could be increased.

Acknowledgment

The authors acknowledge to the National Office for Research and Innovation (NKTH), assistance granted under the Hungarian-Indian (KTIA-DST) R&D&I Program (TÉT_13_DST-1-2014-0003).

References

1. A. López, I. de Marco, B. M. Caballero, M. F. Laresgoiti, A. Adrados, A. Torres, Waste Manage, **31**, 1973 (2011)
2. S. S. Lam, A. D. Russell, C. L. Lee, H. A. Chase, Fuel, **92**, 327 (2012)
3. R. Aguado, G. Elordi, A. Arrizabalaga, M. Artetxe, J. Bilbao, M. Olazar, Chem. Eng. J. **254**, 357 (2014)
4. M. K. Harder, O.T. Forton, J. Anal. App. Pyrol., **79**, 387 (2007)
5. P. T. Williams, S. Besler, D. T. Taylor, Fuel, **69**, 1474 (1990)
6. http://www.idis2.com/ (downloaded at 10 Nov. 2015)
7. http://www.gov.scot/Publications/2013/12/9124/9 (downloaded at 10 Nov. 2015)
8. http://www.nist.gov/lightweighting/ncalfeature.cfm (downloaded at 10 Nov.2015)
9. S. S. Lam, A. D. Russell, C. L. Lee, S. K. Lam, H. A. Chase, Intern. Journal of Hydrogen Energ., **37**, 5011 (2012)
10. J. H. Chen, K. S. Chen, L. Y. Tong, J. of Hazard. Mat., **84**, 43 (2001)
11. J. D. Martínez, J. Rodríguez-Fernández, J. Sánchez-Valdepeñas, R. Murillo, T. García, Fuel, **115**, 490 (2014)
12. B. Plage, A-D. Berg, S. Luhn, Foren. Science Intern., **177**, 146 (2008)
13. P. Kusch, V. Obst, D. Schroeder-Obst, W. Fink, G. Knupp, J. Steinhaus Eng. Failure Analysis, **35**, 114 (2013)
14. J. Zięba-Palus, G. Zadora, J. M. Milczarek, P. Kościelniak, J. of Chrom. A, **1179**, 41 (2008)

MATEC Web of Conferences **49**, 05003 (2016)
DOI: 10.1051/matecconf/20164905003

The Response of Tree-Ring Growth to Climate at Upper Timberline of Southern Aspect of Mt. Taibai

QIN Jin, BAI Hongying [a], SU Kai, LI Shuheng, WANG Jun

College of urban and environment science, Northwest University, Xi'an 710127, China

Abstract. In recent years, the impact of climate change on vegetation in Qinling mountainous area has already been authenticated by numerous investigations, nevertheless, as the major ridge of Qinling Mountains as well as national natural conservation reserve, the ecology response of Mt. Taibai sub-alpine vegetation to climate change has not yet gained enough public attention.In this study, in accordance with the method of dendrochronology, response analysis was carried out to contrast characteristic parameters of tree-ring width chronologies for *Larix chinensis* from different elevations as well as their response pattern to climate change. The result showed that, Mean sensitivity, standard deviation and variance in first eigenvector are increasing with the rise of elevation, but the correlation coefficients (R1, R2, R3) were decreasing which indicated that the strength of the tree's common or relative response to environment was decreasing with altitude. Precipitation had stronger correlation with the tree-ring radial growth than air temperature in both of the sites, during the growing season, trees in lower altitude had better correlation with temperature than in higher altitude, thus showing the different response to climate between the two different sites.

1 Introduction

Tree-ring dating, also known as dendrochronology has been an important climate investigation method all over the world during the past century, climatologists regard it as a promising means to study the climate change in the past due to its high resolution of time and precise dating method [1].

Through analyzing the tree-ring data could help us to understand climate behaviors and its background mechanisms in the specific region and predict the future [2-6]. In the extreme environments between the high mountains and sub-arctic zones, the climate of the forest margin is the main limiting factor of tree growth [7], nevertheless, previous research has shown that the relationship between tree-ring width and climate is complex [8,9].

Qinling mountains, as the demarcation line of northern and southern China, mainly lies in the middle of Shaanxi province，northwest China. The *Larix chinensis* in the southern slope of its main peak Taibai mountain conveyed relatively clear signals of the climate change, they are vulnerable and sensitive if the climate continues to change because they can grow healthily only under specific conditions. As a result, studies in this region to obtain response variation information of the tree-width to climate is becoming more critical.Wealth of information was included in the samples of *Larix chinesis* located in the upper slope position and forest margin, so samples in these areas were suitable for doing analysis of annual rings[10]and a 30 to 40 years cycle was found in temperature changes in the region [11], furthermore,tree-ring width chronology has been built from *Larix chinensis* at the divide sampling site in the mid-Qinling Mountains from 1814 to 2003 [12]. However, different elevations in this area had a subsequent various consequences on the tree-ring width and the radial growth of the plant,the terrain and local climate characteristics of Qinling Mountains is complicated and changeable,most studies were targeted at the northern aspect of Mt.Taibai,but the relationship between tree growths and climate at the southern aspect around the forest margin is still unknown, further investigation is required.

In this paper, the two sample sites of the tree coresare close to each other (3100m, 3200m) so the distinctions of the response to climate under such a situation are what we mainly care about.

2 Materials and methods

2.1 Site description and tree-ring material

The tree samples were collected in the southeastern slope of Taibai Mountain around the forest margin in summer of 2012. It is located at the middle western part (N33°49′~34°10′,E 107°19′~107°58′) of the whole Qinling Mountains, also, on the edge of the subtropical monsoon climate area which has a huge barrier function on the warm air from the south in summer and the cold air from the north in winter. The habitat of *Larix chinensis* only covered the elevation at thee range of

[a] Corresponding author: Hongyingbai@163.com

2600-3500m,the annual mean temperature of this height is - 2 ℃ ~ - 1 ℃ , annual total precipitation is about 800~900 mm.(Figure 1)

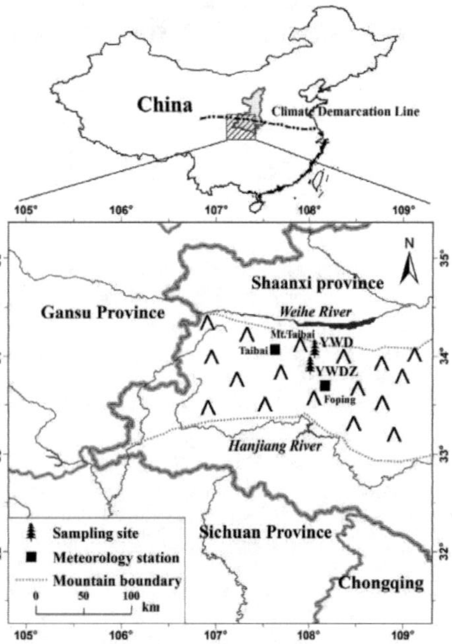

Figure 1. The major characteristics of the trees in two sample site

According to the ITRDB standard, increment cores were collected from 30 living trees in 2 individual 20m×20m plots, choose 15 trees per site at the height of 1.3m (DBH) for extraction, named as YWD (upper) and YWDZ (lower) 2 cores per tree were perpendicularly extracted with the blue Sweden increment borer. The major characteristics of the two sites are listed in Table 1.It is obvious that the trees in YWDZ had a better living condition than YWD in all aspects given.

Table 1. The major characteristics of the trees in two sample sites

Sample site	range of height(m)	range of diameter(cm)	range of crown density(%)
YWD	6-15	18.15-31.21	30-65
YWDZ	13-20	22.92-38.69	70-75

Tree-ring samples were treated following standard practices [13]. All cores were dried, glued, fixed and mounted. After cross-dating, each individual ring was identified with an accurate calendar year, and then each annual ring width was measured within 0.01 mm by manipulating the professional facility LINTAB 5. The quality control of cross-dating was operated by using COFECHA program [14]. Some of the samples found to be significantly hard to cross-date were abandoned from the development of the chronologies, general information of the sample sites is summarized in Table 2.

Table 2. The general condition of the sample site

Sample site	North longitude	East latitude	Elevation	Sample cores	Length
YWDZ1	107°46′43″	33°59′3″	3107	10	100
YWDZ2	107°46′44″	33°55′8″	3093	10	116
YWDZ3	107°46′45″	33°56′	3122	10	95
YWD1	107°46′47″	33°56′	3200	10	100
YWD2	107°46′7.5″	33°56′1″	3213	10	155
YWD3	107°46′6.6″	33°56′1″	3212	10	259

2.2 Chronology development

The individual ring-width measurement series were double detrended and standardized to ring-width indices by using the ARSTAN program [15]. Undesirable growth trends, related to age and stand dynamics but unrelated to climatic variations, were removed from each series during the detrending process. To conserve the maximum common signal at the lowest frequency possible, each ring-width measurement series was standardized conservatively by fitting a negative exponential or straight line. All individual index series were combined into a single chronology by computing a bi-weight robust mean. In further analysis, the Residual version (RES) of the chro-nology was used since it provides relatively higher response signals and more high frequency signals than the STD chronology in our research. In additionn, a lot of works had been done in the past was using RES chronology for investigating the differences in responses to climate.

The statistical characteristics of the RES chronol-ogy in different elevations are listed in Table 3.

Table 3. Statistical features of the YWD and YWDZ residual chronologies

Statistical item	YWD	YWDZ
Mean sensitivity	0.3043	0.2055
standard deviation	0.2803	0.1841
First order auto-correlation	0.0793	0.0128
Variance in first eigenvector(%)	55.70%	54.99%
Signal-to-noise ratio	2.711	2.726
R1	0.493	0.509
R2	0.475	0.476
R3	0.537	0.594

R1 is the average correlation coefficient between all cores, R2 is the correlation coefficient between different trees, R3 is the correlation coefficient between all samples from the same tree.

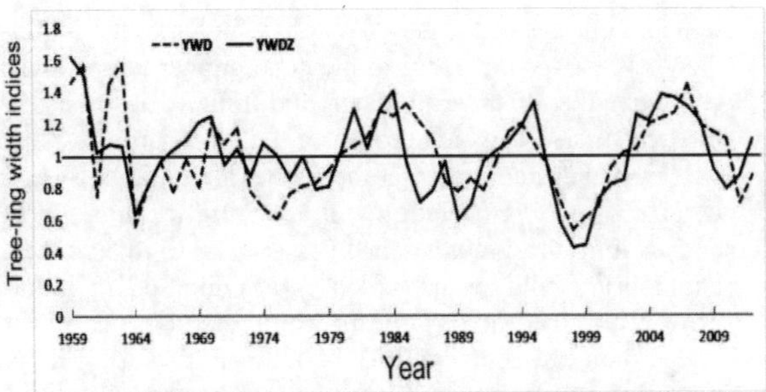

Figure 2. The curves of the RES chronologies of YWD and YWDZ in common interval.

The mean sensitivity (M. S.) and standard deviation (S. D.) of YWD is higher than YWDZ, which means the year-to-year variability of plant growth was increasing with elevation. In conclusion, the *Larix chinensis* from the site of YWD are more sensitive to climate than the samples from YWDZ.

The variance of the first eigenvalue (PC1) of each chronology, expressing the degree of common signals, also increases with altitude. Therefore, the consistency of the tree's response to climate are directly linked to altitude even if there is subtle level difference in elevation.

Nevertheless, the signal to noise ratio, also as SNR which is the ratio of climate signal to non-climate signal, is higher in YWDZ (2.726) than YWD (2.711) but they were almost the same and the correlation coefficients (R1, R2, R3) in RES chronology of both sites indicated that the the trees in YWD and YWDZ both contains high signal.Hence, both sites are suitable for dendrochronology study.

We can infer from the curves of the public interval RES chronologies from the YWD and YWDZ in their common interval (Figure 2) that the chronologies have very similar variations, such as the dumps during 1913, 1963 and 1998, the humps during 1930 and 1945, since they were extracted from the same climate region. In other words, similar variations in the chronologies suggest the tree-ring width at different elevations is affected by similar climate factors.

2.3 Meteorological data

In order to understand the varied response of trees to climate at such different elevations and find which climate factor is the main one responsible for the difference, the correlation between the 2 chronologies to both precipitation and temperature in different months was calculated.

The meteorological data used in this analysis are from the Taibai weather station and Foping weather station since 1959 (Figure 3).The Taibai station lies in the northwestern side of the sample sites and the Foping station lies in the southern side of the sample sites, their major information is summarized in Table 4, they are both close to the sample sites and located in the same climate region (Table 4), so we calculated the annual mean temperature and annual mean precipitation of the two stations in different months for analysis.

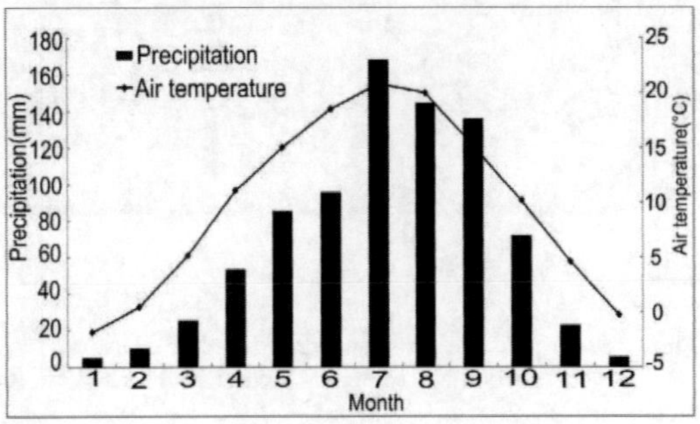

Figure 3. Calculated monthly mean air temperature and precipitation of the Taibai and Foping station

Table 4. The information about the meteorological stations

Stations	Elevation	North longitude	East latitude
Taibai	1543	34°5' 24"	107°17' 59"
Foping	1087	33°32' 59"	108°0' 0"

It is conspicuous that the air temperature and the precipitation in this area were increasing and dropping at the same period which is related to characteristics the temperate monsoon climate, the temperature and precipitation reached their peak in July. (Figure 3).

3 Results and discussion

It is critical to measure if the chronology could reflect the climate signal or not by doing correlation analysis between the chronology and the meteorological data when doing climate research with tree-ring width [16]. To calculate the correlations between the 2RES chronologies and the meteorological data since 1959 and find the divergence of their response to climate, we choose Dendroclim 2002 and SPSS as the software for statistical analysis..Because of the lagging effect, we select the mean monthly precipitation and mean monthly air temperature from October of previous year to October of current year for study (Figure 4, Figure 5).

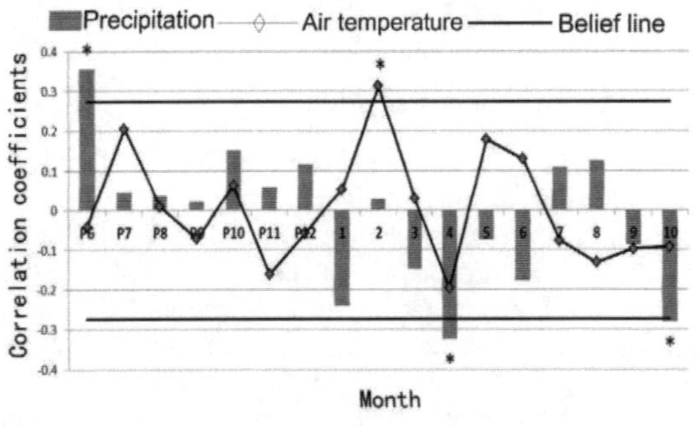

Figure 4. Response function analysis for the relationship between RES chronology of and climate factors in YWD

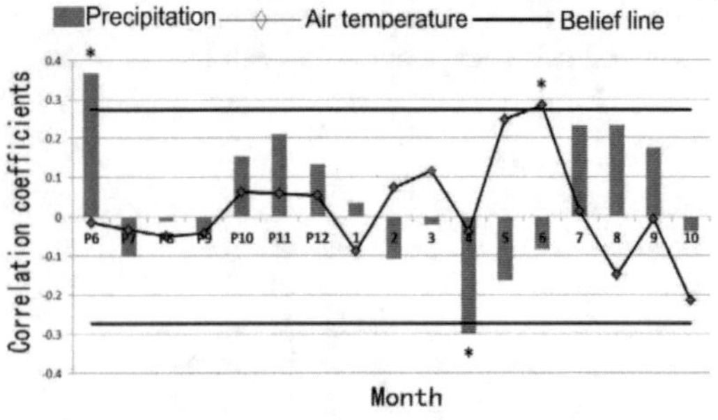

Figure 5. Response function analysis for the relationship between RES chronology of tree-ring width indices and climate factors in YWDZ

3.1 Common characters

It is obvious that the two graphs reflected some similar characteristics.

To precipitation, both of the two sites' tree-ring widths had notable positive correlation with precipitation in June of previous year, it means the precipitation in previous year had stronger positive effect on the radial growth of the trees even to current year, this phenomenon is directly linked to the "lagging effect", which proved that the *Larix chinensis* had notable "lagging effect" with precipitation; Another conspicuous resemble character is that the tree-ring width of the two sites all respond notably positively to precipitation in April in current year, this may due to the special environmental condition at that time, during the days in April the ice of the mountains began to melt, the trees had enough water for grow but not enough temperature and light, more precipitation could limit the time of sunlight and warmth then turned out to be the significantly slowing down the speed of tree-ring radial growth.

In summer, from April to July better condition of temperature can help to increase the speed of photosynthesis, in August when temperature began to fall and the grow season is at the end of the corner, high temperature would cause higher evaporation then it has negative correlation with the temperature again.

In conclusion, it can be inferred from the data and graphs that the precipitation had stronger correlation with the radial growth than air temperature in both of the sites,this proved them to be in the same climate region. However, there still exists some significant differences in correlation strength between the two chronologies.

3.2 Different characteristics

Despite the similarities in the two chronologies, there are some distinctions can't be ignored.

After comparison of the two sample sites in Table 1, we can tell that the total living status in YWDZ is better than YWD, his can be described as the result of elevation, the higher the elevation is , the worse environment it will be.

It can be revealed from Figure 4 and Figure 5 that YWD had significant positive correlation with temperature in February, because in spring when temperature began to rise, YWD had better water supply than YWDZ for it is close to the peak, higher temperature can make the ice melting faster, and initially absorbed by the trees in YWD which promoted the radial growth.

In winter during October, the precipitation had notable negative correlation with tree-ring width in YWD, at that time the growth season is over, the precipitation could bring cold water which is a limitation in radial growth and the photosynthesis would also be influenced under such low temperature, the temperature in YWD is relatively colder than YWDZ, as a result, YWD had rather significant negative correlation with precipitation than YWDZ at that time. There is a remarkable peak in temperature in the Fig. 5 showed that in the growth season in YWDZ had very good correlation with temperature than YWD, this can be explained as the YWD' s air temperature is always lower than YWDZ, in growing season of June the temperature in YWDZ is more suitable for radial growth than the trees in YWD.

4 Conclusions

(1) The living condition of the *Larix chinensis* in YWDZ is better than YWD, in other words, the *Larix Chinensis* in YWDZ are healthier than YWD, the environment in YWDZ are more suitable for the *Larix Chinensis* to habitat.

(2) Values of statistical features of the YWD and YWDZ residual chronologies are high enough for dendrochronology study,while the chronology of YWD contains more information of environment change than YWDZ.

(3) The precipitation had stronger correlation with the radial growth than air temperature in both of the sites, most importantly, precipitation in June of previous year had significant lagging effect with width indices , that is to say, precipitation is the limiting factor of the radial growth for both of the sites. In the growing season in YWDZ had very good correlation with temperature than YWD.Their significant differences had physiological meaning.

However,the work of climate reconstruction during the last 100 years between the two sites at Mt.Taibai haven't been done yet ,in order to have a better understanding of the climate change around this elevation and explain why the *Abies fargesii* at the timberline could response differently to climate change.Our next study will put an emphasis on climate reconstruction in this region to figure out the historical reasons for the phenomenon found above.

Acknowledgments

We would like to thank Dr. Bao guang from Baoji university of arts and sciences and Dr.Li gang from Northwest university for their great help in this article.

References

1. Fritts, H. C. (eds), London: Academic Press, (1976)

2. Briffa, K. R., Jones, P. D., Schweigruber, F. H., Osborn, T. J., Nat., **393**, 450–454 (1998)
3. Frank, D., Wilson, R., Esper, J., Boreas, **34**, 498–505 (2005)
4. Rutherford, S., Mann, M. E., Osborn, T. J., Bradley, R. S., Briffa, K. R., Hughes, M. K., Jones. P. D., Journal of Climate, **18**, 2308–2329 (2005)
5. Esper, J., Frank, D., Büntgen, U., Verstege, A., Luterbacher, J., Xoplaki, E., Geophysical Research Letters, DOI:10.1029/ 2007GL030844 (2007)
6. Mann, M. E., Zhang, Z. H., Hughes, M. K., Bradley, R. S., Miller, S. K., Rutherford, S., Ni, F. B., Proceedings of the National Academy of Sciences of the United States of America, **105**, 13, 252–13, 257 (2008)
7. Tessier, L., Guibal, F., Schweingruber, F. H., Climatic Change, **36**, 499–517 (1997)
8. Buckley, B. M., Cook, E. R., Peterson, M. J., Barbetti, M., Australia. Climatic Change, **36**, 477–498 (1997)
9. Splechtna, B. E., Dobry, J., Klinka, K., Annals of Forest Science, **57**, 89–100 (2000)
10. Kang, Y. X., Liu, J. H. Dai, S. F., He, X. J., Journey of northwest university of A&F (Nature science edition), **38**, 12, 141-147 (2010)
11. Dai, J. H., Shao, X. M., Cui, H. T., Ge, Q. S., Liu, H. Y., Tang, Z. R., Quaternary Science. **23**, 428-435 (in Chinese with English abstract) (2003)
12. Liu, Y., Liu, N., Song, H. M., Cai, Q. F., Bao, G., Wang, W. P, Advances In Climate Change Research, **5**, 5, 260-265 (2009)
13. Stokes, M. A., Smiley, T. L., The University of Arizona Press,Tucson (1996)
14. Holmes, R. L.,Tree-ring Bulletin **43**, 69–78 (1983)
15. Cook, E. R., D, University of Arizona, Tucson (1985)
16. Fritts, H. C., The University of Arizona Press, Tucson, USA (1991)

MATEC Web of Conferences **49**, 07001 (2016)
DOI: 10.1051/matecconf/20164907001

Prediction of Bacterial Virulent Proteins with Composition Moment Vector Feature Encoding Method

Murat Gök [1,a], Deniz Herand [2]

[1] *Department of Computer Engineering, Yalova University, Yalova, TURKEY*
[2] *Department of Industrial Engineering, Turkish German University, Istanbul, TURKEY*

Abstract. Prediction of bacterial virulent proteins is critical for vaccine development and understanding of virulence mechanisms in pathogens. For this purpose, a number of feature encoding methods based on sequences and evolutionary information of a given protein have been proposed and applied with some classifier algorithms so far. In this paper, we performed composition moment vector (CMV), which includes information about both composition and position of amino acid in the protein sequence to predict bacterial virulent proteins. The tests were validated in three different independent datasets. Experimental results show that CMV feature encoding method leads to better classification performance in terms of accuracy, sensitivity, f-measure and the Matthews correlation coefficient (MCC) scores on diverse classifiers.

1 Introduction

Virulence is the degree of pathogenicity, which is ability to cause disease, within a group or species of bacteria to invade the tissues of the host. The pathogenicity of an organism is determined by its virulence factors that is typically proteins coded for by genes in chromosomal DNA or plasmids [1]. Identification of virulent proteins that cause drug resistant varieties of various bacterial pathogen evoke the design of peptide based vaccine and drug. Bacterial virulent proteins can be classified from the point of mechanisms of virulence such as adhesion, colonization, invasion, immune response inhibitors and bacterial toxins. Many bacteria adhere to the host cells. This class of proteins includes fimbria and pili in Escherichia coli, Vibrio cholerae, Pseudomonas aeruginosa and Neisseria species [2]. Some virulent bacteria such as helicobacter pylori, which survives in the acidic milieu of the human stomach, produce special proteins that allow them to colonize parts of the host body. Many bacteria produce virulence factors that disrupt the host cell membranes or stimulate their own endocytosis or macro pinocytosis into host cells to allow and facilitate the bacteria to enter host cells [3]. Immune response inhibitors are another class of virulent proteins in some bacteria such as streptococcus pneumoniae that inhibit the host's immune system defenses by producing proteins that bind host antibodies. Bacterial toxins that are made by bacteria that poison host cells and cause tissue damage are another common virulence factors.

Although many bacteria genomes, more than 6000, were sequenced, relatively limited number of virulent proteins were discovered. In the literature, there are two approaches to predict virulent proteins: similarity search and machine learning methods. BLAST [5] and PSI-BLAST [6] are typical examples for the former approach. Machine learning approaches are widely used for the problem. In [7], the authors proposed a neural network-based prediction of virulence factors. Also, 2-gram compositions and the higher order dipeptide composition has been applied with an ensemble of SVM. In [2] the authors have presented a method based on ensemble of classifiers for virulent proteins prediction where the features are extracted directly from the amino acid sequence and from the evolutionary information of a given protein.

The aim of this paper is to apply composition moment vector (CMV), which includes information about both composition and position of amino acid in the protein sequence for prediction of bacterial virulent proteins with several machine learning algorithms.

This paper is organized as follows. In Sect. 2, datasets, feature extraction and classification procedures are explained. In Sect. 3, CMV is evaluated and its performance is obtained according to several classifiers. Finally, Sect. 4 concludes the paper.

2 Material and methods

2.1 Dataset

We conducted our tests on two-to-date datasets: Adhensis (ADH) and Independent (IND) datasets [2, 7]. ADH dataset consists of 469 adhesins and 703 nonadhesins proteins (including several archaebacterial, viral, and

[a] Corresponding author: murat.gok@yalova.edu.tr

yeast non-virulent proteins). IND dataset consists of 367 SWISS-PROT sequences (181 virulent and 186 nonvirulent protein sequences).

2.2 Composition moment vector

Due to the fact that in vitro prediction methods for a protein classification task are time consuming and labour expensive, computational methods based on pattern recognition / machine learning algorithms are used.

A complete pattern recognition system, as shown in Figure 1, consists of a sensor that gathers the observations to be classified, the preprocessing module that removes noise, normalizes the pattern, a feature encoding / selection mechanism that encodes / selects numeric information from the pattern and a classification algorithm that implements the classification task [8].

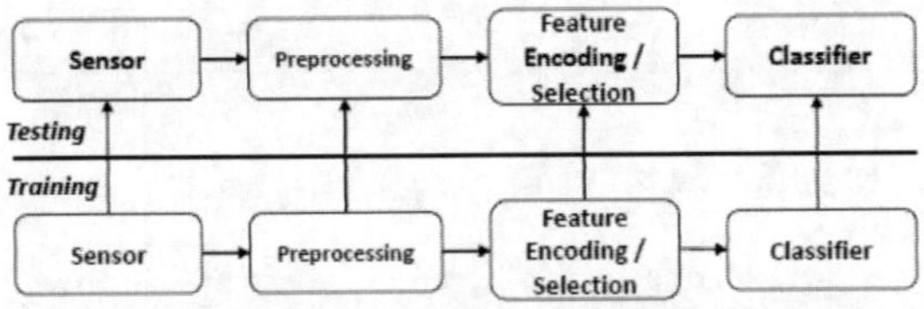

Figure 1. The architecture of a pattern recognition system.

Feature encoding process defines a mapping from the original representation space into a new space where the classes are more easily separable. The goal of feature encoding is to identify the pattern data to the classifier algorithms as much as possible. CMV feature encoding method takes into account amino acid composition and their position information in a sequence. That is, CMV, which includes information about both composition and position of amino acid in the sequence as well as functional relation with the structure content, i.e. there must not be two or more primary amino acid sequences that would have different structure content but the same composition moment vector. Moreover, since it provides information about each AA in the primary sequence, it gives a more comprehensive description of the sequence than other measures [9].

2.3 Classifier algorithms

For classification, we used five types of classifier algorithms: k-Nearest Neighbor (k-NN), Naïve Bayes, Random Forest, C 4.5 and Ada Boost.

The underlying mechanism of k-NN algorithm is quite straightforward that needs no specific training phase. The only things needed are reference data points for both classes (clients, impostors). An unknown (test) data point y is then attributed the same class label as the label of the majority of its k nearest (reference) neighbors. To find these k nearest neighbors the Euclidean distance between the test point and all the reference points is calculated, the obtained distances are ranked in ascending order and the reference points corresponding to the k smallest Euclidean distances are taken. This exhaustive distance calculation step during the test phase leads rapidly to important computing times, which is the major drawback of k-NN [10].

Naïve Bayes is an effective and basic classification algorithm that assumes the feature variables to be independent from each other given the outcome. This assumption simplifies the calculation of conditional probabilities.

With Naïve Bayes algorithm, given a sample, si, the probability of each class, cj, is calculated as in Eq. 1. High probability determined related sample's class.

$$P(c_j \mid s_i) = \frac{P(s_i \mid c_j) \cdot P(c_j)}{P(s_i)} \tag{1}$$

Given a bacterial protein sequence, described by its feature vector $s_i = (s_1, s_2, .., s_n)$, we are looking for a class c_j that maximizes the likelihood $P(s_i \mid c_j) = P(s_1, s_2, .., s_n \mid c_j)$. Thus, each misclassification error, $P(s_i \mid c_j)$, is accounted for s_i with Eq. 1. s_i belongs to the class that provides minimum misclassification error.

The random forest classifier which is a way to improve the performance of Decision Tree consists of a combination of tree classifiers where each classifier is generated using a random vector sampled independently from the input vector, and each tree casts a unit vote for the most popular class to classify an input vector [11]. The random forest classifier used for this study consists of fusing randomly selected features or a combination of features at each node to grow a tree. Design of a decision tree required the choice of an attribute selection measure and a pruning method. There are many approaches to the selection of attributes used for decision tree. Information Gain Ratio criterion [12] and the Gini Index [11] are the main attribute selection measures in decision tree induction. The random forest classifier uses the Gini Index as an attribute selection measure, which measures the impurity of an attribute with respect to the classes [13].

C4.5 algorithm uses a divide-and-conquer approach to learn decision trees which are a very effective method of supervised learning. A decision tree aims the partition of a dataset into groups as homogeneous as possible in terms of the variable to be predicted. It takes as input a set of classified data, and outputs a tree that resembles to an orientation diagram where each end node (leaf) is a decision (a class) and each non- final node (internal) represents a test. Each leaf represents the decision of belonging to a class of data verifying all tests path from the root to the leaf. The data is sorted at every node of the tree in order to determine the best splitting attribute. It uses gain ratio impurity method to evaluate the splitting attribute. Decision trees are built in C4.5 by using a set of training data. At each node of the tree, C4.5 chooses one attribute of the data that most effectively splits its set of samples into subsets enriched in one class or the other. Its criterion is the normalized information gain (difference in entropy) that results from choosing an attribute for splitting the data. The attribute with the highest normalized information gain is chosen to make the decision [14].

The AdaBoost (adaptive boosting) algorithm is an ensemble of classifier algorithms for generating a strong classier out of a set of weak classifiers [15]. In the process of using weak classification repeatedly, different distributions of sample sets should be used, or a weighted sample set, to train the simple classifiers. The weight of a sample set is calculated as in Eq. 2.

$$d^t = \{d_1^t, d_1^t, ..., d_1^t\}, \sum_{n=1}^{N} d_n^t = 1, d_n^t \geq 0 \quad (2)$$

In each training, the misclassified samples will have larger weight during the next training. In general, the samples closest to the decision-making boundary will be easily misclassified. Therefore, after several iterations, these samples assume the greatest weights. As a result, if there are enough training samples and classification errors, we can obtain a stronger classifier through the AdaBoost algorithm [16]. In this study, we have ensemble Decision Stump with AdaBoost classifier.

3 Results and discussion

3.1 Experimental setup

The performances of the classifiers were evaluated by means of accuracy, F-score and the Matthews correlation coefficient (MCC) performance metrics. True positives (TP), true negatives (TN), false positives (FP) and false negatives (FN) values are obtained via confusion matrix. Acc is a widely used measure to determine class discrimination ability, and it is calculated as:

$$accuracy = \frac{TP + TN}{TP + FP + TN + FN} \quad (3)$$

Specificity is the ratio of TP prediction and it is calculated as:

$$sensitivity = \frac{TP}{TP + FN} \quad (4)$$

Matthew's correlation coefficient (MCC), which is used as a measure of the quality of binary classifications, takes into account true and false positives and negatives.

$$MCC = \frac{TP \times TN - FP \times FN}{\sqrt{(TP + FP)(TP + FN)(TN + FP)(TN + FN)}} \quad (5)$$

F-score is a measure of a test's accuracy determining accuracy accounting for both precision and for sensitivity from confusion matrix. F-score accounted as shown in Eq. 6.

$$F - score = 2 \times \frac{precision \times sensitivity}{precision + sensitivity} \quad (6)$$

MCC and F-score takes values between (−1, 1). The higher MCC and F-score get, the more realistic performance classifier system gives [17].

3.2 Performance of CMV

10-fold cross validation (10-fold CV) testing scheme is applied to evaluate the performance of the methods in terms of accuracy, F-score and averaged over 10 experiments on ADH and IND-1 datasets. In a 10-fold CV the training samples are randomly partitioned into 10 equal sized folds with similar class distributions. Each fold in turn is then used as test data for the classifier generated from the remaining nine folds [11]. Having completed the procedures above, the average accuracy, F-score and MCC values of the each method over these 10 turns are obtained, as shown Table 1 and Table 2.

Table 1. Accuracy, F-Score and MCC Performances on ADH Dataset.

Classifiers	Accuracy (%)	Sensitivity (%)	F-score	MCC
k-NN	92.66	90	0.90	0.85
Random Forest	91.21	85.90	0.89	0.82
Naive Bayes	85.07	77.60	0.81	0.69
C4.5	84.73	79.50	0.81	0.69
Ada Boost	83.87	72.70	0.78	0.66

The results report that CMV encoding gives the best result with k-NN algorithm with accuracy value of 92.66 %, sensitivity value of 90 % F-score value of 0.93.

Table 2. Accuracy, F-Score and MCC Performances on IND Dataset.

Classifiers	Accuracy (%)	Sensitivity (%)	F-score	MCC
Rotation Forest	71.12	70.70	0.71	0.42
Naive Bayes	69.21	61.30	0.66	0.39
Ada Boost	68.66	65.50	0.66	0.37
C4.5	63.21	54.70	0.60	0.27
k-NN	62.67	58.60	0.61	0.25

The results in Table 2 points out that Rotation Forest algorithm has obtained the best result for accuracy and MCC values with the value of 71.12 % and 0.42, respectively. Note that kNN algorithm obtained the worst performance scores from the point of accuracy and MCC values among learning algorithms.

4 Conclusion

The problem addressed in this paper is to predict bacterial virulent proteins with several machine learning algorithms using CMV feature encoding method on two virulent protein datasets. We performed an experimental comparison of five classifiers: k-NN, Random Forest, Naïve Bayes, C4.5 and Ada Boost. Experimental results show that CMV method is fit for this problem with different classifiers. Next studies will involve new feature encoding methods which are free of sequence size using ensemble of classifiers. Furthermore, due to the fact that publicly accessible web servers are more practical to study and to develop more useful models or predictors, we will study on developing a web server to predict bacterial virulent proteins.

Acknowledgment

This work was supported by Yalova University, BAP Project (Grant 2015/BAP/103).

References

1. Morens, D. M., Folkers, G. K., and A. S. Fauci, Nature, **430**, 242-249 (2004)
2. Nanni, L., Lumini, A., Gupta, D., and Garg, A. IEEE/ACM Transactions on Computational Biology and Bioinformatics (TCBB), **9**, 467-475 (2012)
3. YashRoy, R. C. Indian J Med Res, **126**, 558-566 (2007)
4. Brogden, K. A., Roth, J. A., Stanton, T. B. *et al.* Virulence Mechanisms of Bacterial Pathogens, (third ed.). ASM Press (2000)
5. Altschul, S. F., Gish, W. Miller, W. *et al.* J. Molecular Biology, **215**, 403-410 (1990)
6. Altschul, S. F., Madden, T. L., Schaffer, A. A. *et al.* Nucleic Acids Research, **25**, 3389-3402 (1997)
7. Sachdeva, G., Kumar, K. Jain, P., and Ramachandran, S. Bioinformatics, **21**, 483-91 (2005)
8. Zheng N., Xue J., Statistical Learning and Pattern Analysis for Image and Video Processing, Springer-Verlag, 3-4 (2009)
9. Ruan, J., Wang, K., Yang, J. *et al.* Artificial Intelligence in Medicine, **35**, 19-35 (2005)
10. Verlinde, P., and Cholet, G. Int. Conf. Audio and Video-Based Biometric Person Authentication (AVBPA), 188-193 (1999)
11. Breiman, L. Technical Report 567, Statistics Department, University of California, Berkeley, ftp://ftp.stat.berkeley.edu/pub/users/breiman (1999)
12. Quinlan, J. R., C4.5: Programs for Machine Learning, San Mateo, CA: Morgan Kaufmann (1993)
13. Pal, M. International Journal of Remote Sensing, **26**, 1, 217-222 (2005)
14. Hssina, B, Merbouha, A., Ezzikouri, H., Erritali, M. Int. J. of Adv. Comp. Sci. and App., **4**, 13-19 (2014)
15. Freund, Y. and Shapire, R. Proceedings of the Second European Conference on Computational Learning Theory, 23-37 (1995)
16. Zheng, N. and Xue, J. Statistical learning and pattern analysis for image and video processing, Springer-Verlag London, 172-173 (2009)
17. Gök, M. International Journal of Systems Science, **46**, 1108-1112 (2015)

MATEC Web of Conferences **49**, 07002 (2016)
DOI: 10.1051/matecconf/20164907002

Molecular Studies on Preproinsulin Gene

Sarah Sabir [a], Hooria Younas

Kinnaird College for Women 93-Jail Road, Lahore Pakistan

Abstract. Insulin plays an important role in maintaining the blood glucose level of the body. The β -cells of pancreas produce insulin in the form of precursor that is preproinsulin. The gene of preproinsulin provides an interesting system for addressing question related to molecular evolution. Recombinant DNA technology has made it possible to isolate and sequence the chromosomal genes coding for unique protein products. Although preproinsulin of various organism has been isolated and cloned, but there is no report from buffalo (Bubalus bubalis) that is our major livestock. The genomic DNA of buffalo was isolated using Laura-Lee-Boodram method. The part of preproinsulin gene (596bp and 520bp) using BPPI-UPS and bpiful_F as forward and BC1-C as reverse primer was amplified. Cloning of amplified fragments of gene were performed in pCR 2.1 vector. Positive clones were screened on the basis of blue white selection. The band obtained on 596bp and 520bp after colony PCR confirmed the successful cloning of preproinsulin gene in pCR 2.1 vector.

1 Introduction

Diabetes mellitus is a metabolic disease in which blood glucose level of the person is high because the body does not produce insulin or the cells do not respond to the insulin which is produced in the body [1]. The main role of insulin in liver, muscles and fat tissues is to control the sugar level, proteins and fat metabolism. The major role of insulin is to regulate blood glucose level, to promote protein synthesis and prevent its decomposition [2]. High blood glucose level stimulates the β- cells of pancreas to release insulin, which allows the glucose to enter the body's cell and it is used as a fuel for energy. It maintains the blood sugar level in the range of 80-120mg/dl. If the glucose level goes low that another hormone glucagon is produced by the alpha cells of pancreas in the opposite direction as compared to insulin. The function of glucagon is to release the stored glucose in the bloodstream [2]. Insulin is a small protein of 51 amino acids and molecular weight of about 6000 Da. It is composed of two polypeptide chains; chain A has 21 amino acids and chain B has 30 amino acids [3]. Insulin is synthesized as a single precursor molecule, preproinsulin, which is posttranslationally modified to give the mature hormone [4]. Beta cells release insulin in response to change in glucose level. Type 2 glucose transporters are responsible for the entry of glucose in the cells. The glucokinase modify the glucose molecules by adding phosphate group so that they are trapped in beta cells and is mobilized to ATP. As the level of ATP rises the potassium gated channels close up, preventing potassium ions from being shunted across the cell membrane. The rise in positive charge due to potassium ions causes depolarization, this causes the calcium ion channels to open and transport calcium ions in the cells. The increase in calcium ions trigger the export of insulin storing granules by exocytosis and this results in the transport of insulin out of the cells in the bloodstream[5]. When insulin reaches the liver cells it bind to its receptors present on the cell member of the liver, as it binds a number of different chemical reactions are affected in the liver. Firstly it causes the liver cells to increase the breakdown of glucose and secondly it causes the glucose to be converted to glycogen and stored in the liver cells [5]. The genes and mRNAs that encode preproinsulin have been cloned from a variety of species, mouse [6], human[7], guinea pig[8], chicken [9], dog [10], birds [11], bovine [12] and buffalo proinsulin mRNA (Accession No AB234871.1) [13] have been sequenced. So, to fill the gap preproinsulin from buffalo (Bubalus bubalis).

2 Materials and methods

Blood sample of buffalo (*Bubalus bubalis*) was collected from slaughter house in a vial containing 10μl of EDTA.

Genomic DNA wad isolated from this sample by Laura-Lee-Boodram method [14]. The quality and quantity of genomic DNA was analyzed on 1% agarose gel.

In the present study following primer sets were used to amplify the part of preproinsulin gene.

BPPI-UPS(Forward):
GCTCCTGGAGGCCCTGATGACGGGCCAG

Bpiful_F(Forward):
TCATATGGACCGGCTGCATTCGAGGCTGCCAG.

[a] Corresponding author: sarahrehmat@gmail.com

BC1-C(Reverse):
GCCGCGCTTCTGCGGGGGGCCCTCCACCTCCCGG CGGGC.

Bpi-R(Reverse):
GTTTAAGCTTGTCGGGGGCAGGCCTAGTTACAG

All the primers were dissolved in autoclaved water to make their final concentration 100pmol/µl. The aliquots of these primers were stored at -20°C.

Four PCR reaction mixtures, each of 50 µl were prepared as mentioned in table 3.8. The reaction mixtures were incubated in thermal cycler at 95°C for 5 minutes and were then amplified for 35cycles, each consist of 45 seconds at 95°C, 0:30 seconds at 55°C for annealing and 1:10 at 72°C. The reaction mixtures were incubated of 5 minutes at 72°C for complete extension. PCR product was analyzed on 1% Agarose gel electrophoresis as described previously.

PCR product was analyzed on 1% Agarose gel electrophoresis

Products and the band of PCR was excised and DNA was extracted from gel using commercially available "PureLink® Quick Gel extraction Kit" Invitrogen Company using the protocol provided with kit.

The purified PCR product was ligated into a cloning vector, pCR® 2.1 vector (20ng/µl) using "The Quality TA cloning® Kit".20µl of ligation mixture for each PCR product was prepared in Eppendorf tubes and it was incubated at 14°C in cooling incubator for overnight as mentioned in Table 1.

Table 1. Preparation of Ligation mixture for T/A ligation of preproinsulin gene.

Reagents	PCR Product (BPPI-UPS/BC1-C)	PCR Product (bpiful_F/BC1-C)
5X ligation buffer	4 µl	4 µl
PCR product	12 µl	12 µl
T4 DNA ligase 5U/µl	20µl	20 µl
Total	20µl	20 µl

200µl of competent cells of E.Coli DH5α were added in each sample of 20µl ligation mixture. Negative control contains only 200µl competent cells and Positive control contains 200µl of competent cells and Bpi/pTZ57R/T recombinant construct. Samples were mixed gently and placed on ice for 40 minutes. Screening of positive clones was performed by blue white selection method [15] using LB agar plates containing IPTG, X-Gal and ampicillin.

The ligation of BPPI-UPS and bpiful_F gene into pCR® 2.1 vector was confirmed by colony PCR. Colony PCR reaction mixture was prepared in PCR tubes, each 25µl, as mentioned in table 1. Five PCR reaction mixtures, each of 50 µl were prepared. The reaction mixtures were incubated in thermal cycler at 95°C for 5 minutes and were then amplified for 35cycles, each consist of 45 seconds at 95°C, 0:30 seconds at 55°C for annealing and 1:10 at 72°C. The reaction mixtures were incubated of 5 minutes at 72°C for complete extension. PCR product was analyzed on 1% Agarose gel electrophoresis as described

3 Results and discussions

In eukaryotes insulin is synthesized in the beta cells of pancreas in the form of its precursor, preproinsulin. The preregion of the precursor serves as the hydrophobic signal for the transport of the chain through the membrane of endoplasmic reticulum where the signal sequence is cleaved. Then proinsulin attains its secondary structure and disulfide bond formation takes place and finally converted to insulin. The purpose of the study was to clone the preproinsulin gene of buffalo to highlight the evolutionary aspect if insulin and the species to species variation at the DNA level as well as at protein level.

In this study, Genomic DNA was isolated from buffalo blood by Laura-Lee-Boodram method [14]. Gel electrophoresis was performed to determine the quality and quantity of the DNA extracted using Gel documentation system. The gel (Figure 1) showed bands of DNA. Genomic DNA in the wells and 23 kb band of smallest fragment of genomic DNA showed good isolation of genomic DNA. There is no smear of degrade DNA in the gel that shows that the quality of DNA is good and it can be used for further analysis.

There are certain difficulties in isolation of genomic DNA. The pH of buffer A and B should be set otherwise it will cause problem in isolation. After centrifugation supernatant should be discarded carefully so that pellet is not dislodged. Heating process should also be done carefully to inactivate the nuclease if this step is not done carefully the genomic DNA will be cleaved by the nucleases.

Cow and buffalo are closely related species, so primers were designed from the cow sequence (Accession No. EU518675.1) using a software. In case of cow preproinsulin gene there are three exons and two introns. . Four primers were designed to carry out this study. BPPI-UPS (forward primer) was designed from the upstream region of the preproinsulin gene, it was designed to study the promoter region of the preproinsulin gene. Reverse primer Bpi-R was designed near the stop codon so that whole gene can be amplified. The expected fragment size obtained after amplification with BPPI-UPS and Bpi-R is 1213bp. The fragment amplified from these primers contains three exons two introns start codon and stop codon. There was a possibility that there is a difference in cow and buffalo promoter region so another primer bpiful_F was designed from the first exon of the cow preproinsulin gene. The excepted product size is 1137bp using bpiful_F and Bpi-R as forward and reverse primers respectively. Since there was difficulty in amplifying the fragment larger than 1kb, so another reverse primer BC1-C was designed from the middle region of the gene sequence. Expected size of fragment amplified with BPPI-UPS and BC1-C is 596bp and with bpiful_F and BC1-C is 520 bp.

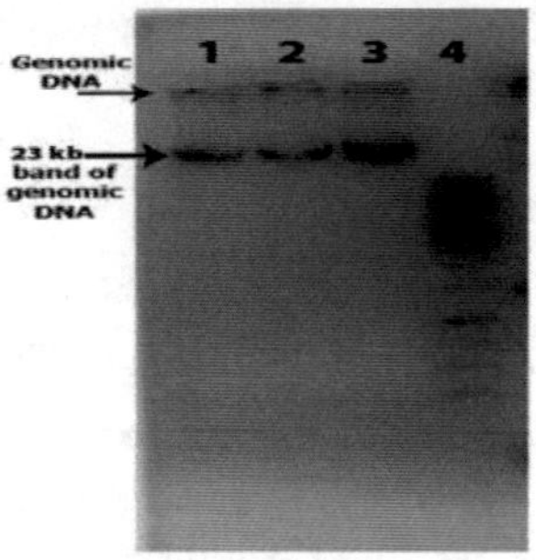

Figure 1. Isolation of genomic DNA.
Lane 1: Buffalo Genomic DNA
Lane 2: Buffalo Genomic DNA
Lane 3: Buffalo Genomic DNA
Lane 4: DNA Ladder

The amplification of preproinsulin gene was a difficult task and many attempts were made to amplify the whole gene of preproinsulin using BPPI-UPS and Bpi-R and bpiful_F and Bpi-R, by varying different conditions of PCR such as annealing temperature, concentration of $MgCl_2$ and the amount of DNA template

Magnesium chloride is used as a co factor for the thermostable *Taq DNA polymerase.* It also stabilizes the double stranded DNA and increases the Tm. Concentration of $MgCl_2$ is important in controlling the specificity of the reaction. Too low $MgCl_2$ concentration results in low yield of PCR product and too high $MgCl_2$ concentration results in non- specific amplifications. $MgCl_2$ concentration should be used in range of 1mM-4mM. That is why different concentration of $MgCl_2$ is used in this study.

Due to larger size fragment of 1213bp there was difficulty in amplification of full preproinsulin gene. So four reaction mixtures were prepared in which BC1-C primer was designed from the middle of the gene sequence, it was used as reverse primer in combination with two forward primers BPPI-UPS and bpiful_F. DNA band at 520bp was observed in lane 1 and lane 2, and 596bp of DNA band was observed in lane 4 and lane 5 on 1% agarose gel.

DNA bands of amplified preproinsulin 596bp (BPPI-UPS/BC1-C) and 520bp (bpiful_F/BC1-C) were cut from gel and gene clean was done, as described previously. The gene was ligated into a cloning vector, pCR® 2.1 vector (20ng/µl) using "The Quality TA cloning® Kit".

Cloning of the PCR product in the vector was done by T/A cloning. When PCR product is amplified using Taq polymerase then it is capable of adding adenosine triphosphate residue to the 3' ends of the double stranded PCR product. Vector is digested with Eco32I restriction enzyme so that the amplified PCR product is cloned in a linearized vector with complementary 3'T overhangs. PCR product with A overhangs is mixed with the vector in high proportion. The complementary T overhangs of a vector and the A overhang of PCR product hybridizes and the nick is sealed by DNA ligase [15].

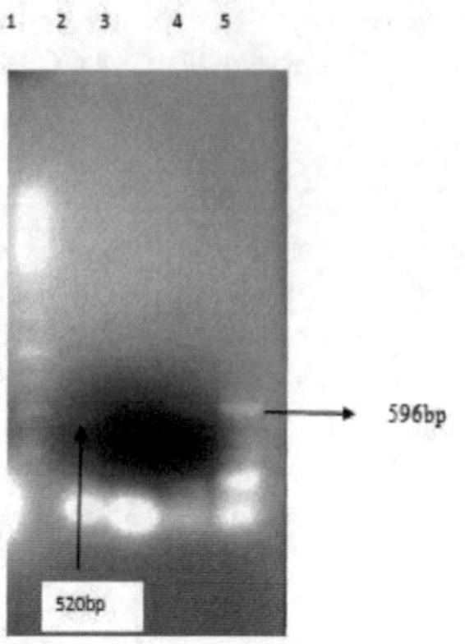

Figure 2. Isolation of PCR amplification of buffalo preproinsulin gene using BPPI-UPS and bpiful_F as forward and BC1-C as reverse primer.
Lane 1: DNA Ladder
Lane 2: PCR amplification of buffalo preproinsulin gene using bpiful_F and BC1-C, 1.5mM MgCl2
Lane 3: PCR amplification of buffalo preproinsulin gene using bpiful_F and BC1-C, 2mM MgCl2
Lane 4: PCR amplification of buffalo preproinsulin gene using BPPI-UPS and BC1-C, 1.5mM MgCl2

(a) (b)

Figure 3.
(a) White colonies of preproinsulin gene amplified by BPPI-UPS and BC1-C
(b) White colonies of preproinsulin gene amplified by bpiful_F and BC1-C

The pCR 2.1 vector contains a fragment of a gene called *LacZ. LacZ* gene codes for the production of an enzyme β-galactosidase. Normally β-galactosidase metabolizes galactose into lactose and glucose. X-Gal is a substrate for galactose, so β-galactosidase converts X-Gal, a colorless modified galactose sugar, into a blue colored product. Within the *LacZα* gene there is a multiple cloning site (MCS) where the PCR product can be cloned. The multiple cloning site (MCS) can be cut with the help of restriction enzyme and the PCR product can be inserted within the *LacZα* gene, this disrupts the function of *LacZ* gene hence no α- peptide is produced. As a result no active β-galactosidase can be produced [16] that cannot convert the x-gal into a blue colored product so white colonies are formed. But if foreign DNA is not ligated in the vector then the enzyme is fully functional and gives rise to bright blue colored colonies [15].

The cloning of preproinsulin gene in pCR 2.1 vector was confirmed by colony PCR using gene specific primers. In colony PCR white bacterial colonies were used as template instead of buffalo genomic DNA. White colonies were picked and mixed with PCR reaction and PCR was run under same conditions.

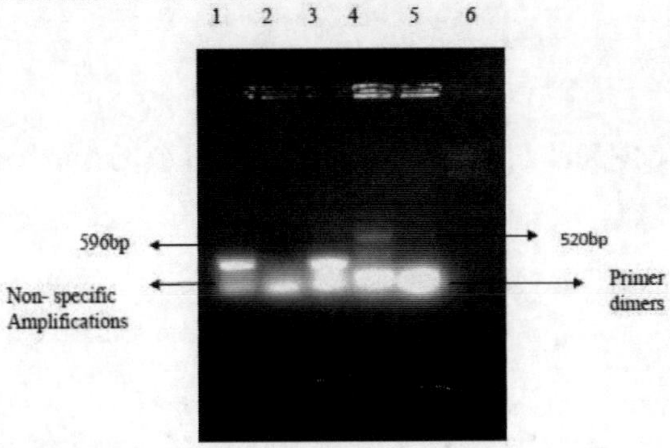

Figure 4. Colony PCR

Lane1: Ampification of BPPI-UPS/BC1-C white colony with BPPI-UPS and BC1-C

Lane 2: Ampification of BPPI-UPS/BC1-C white colony with BPPI-UPS and BC1-C

Lane 3: Ampification of BPPI-UPS/BC1-C white colony with BPPI-UPS and BC1-C

Lane 4: Amplifiaction of bpiful_F/BC1-C white colony with bpiful_F and BC1-C.

Lane 5: Amplifiaction of bpiful_F/BC1-C white colony with bpiful_F and BC1-C.

Lane 6: DNA Ladder

At denaturation step the cells were lysed at 94°C and the recombinant vector was used as template. If the vector contains the insert the primer binds and amplifies it. The presence of 596bp band confirmed that pCR 2.1 vector contained preproinsulin gene amplified by BPPI-UPS and BC1-C. Band at 520bp confirmed that pCR 2.1 vector contains part of preproinsulin gene amplified by bpiful_F and BC1-C primers.

4 Conclusion

Amplification of parts of preproinsulin gene, 596 and 520 bp, was done successfully from buffalo genomic DNA using specific forward and reverse primers which were then successfully cloned in vector pCR 2.1. The cloning was confirmed by colony PCR

5 Recommendation

Sequencing of the parts of preproinsulin gene should be done and comparison of the sequence with other organism should be studied to find out evolutionary aspects.

References

1. K. G. M. M. Alberti, P. Z. Zimmet, Diabetes Medicine, **15**, 539-553 (1998)
2. DF. Steiner, D. Cunningham, L. Spigelman, B. Alen, AAAS, **153**, 697-700 (1969)
3. MJ. Blundel, T. Dodson, E. G. Dodson, M. Vijayan, E. Baker,M. Harding, B. Rimmer, Nature, 224-491 (1969)
4. DF. Steiner, Harvey Lecture, **78**, 191-228 (1983)
5. CR. Kahn, MF. White, JCI, **82**, 4, 1151-1156 (1988)
6. BM. Wentworth, IM. Schaefer, L. VillaKomaroff, JM. Chirgwin, J. Mol. Evol, **23**, 305-12 (1986)
7. SA. Nang, J Biosci **6**,739-55, (1984)
8. SJ. Chan, V. Eposkopou, S. Zeitlin, SK Karathanasis, A. Mackrell, DF. Steiner, *Proc Natl Acad Sci U.S.A*, **81**, 5046-50 (1984)
9. F. Perler, A. Efstratiadis, P. Lomecdico, W. Gilbert, R. Koldner, J. Dodgon, Cell, **20**, 555-66 (1890)
10. S. Kwok, S. Chan. DF. Steiner, J Biol Chem, **258**, 2357-63 (1983)
11. J. Simon, S. Laurent, G. Grolleau, P. Thoraval, D. Soubieux, D. Rasschaert, Mol Phylogenet Evo, **30**, 755-66 (2004)
12. PT. Lomedico, S. Chan, DF. Steiner, J Biol Chem, **252**, 7971-8 (1977)
13. H. Younas, Studies On Recombinant Buffalo (Bubalus Bubalis) Proinsulin And Its Derivatives. PhD thesis, (2009)
14. TL. Kendall, DJ Byerley, R. Dean, Anal Biochem, **195**, 74-6 (1991)
15. D. Chaffin, C. Rubens, Gene, **219**, 91-9 (1998)
16. MY. Zhau, Curr Issues Intest Microbiol, **2**, 1-8 (2000)

MATEC Web of Conferences **49**, 08001 (2016)
DOI: 10.1051/matecconf/20164908001

Comparison between SARS CoV and MERS CoV Using Apriori Algorithm, Decision Tree, SVM

Seongpil Jang [1,a], Seunghwan Lee [1], Seong-Min Choi [1], Junwon Seo [1], Hunseok Choi [1], Taeseon Yoon [2]

[1] *Natural Science, Hankuk Academy of Foreign Studies, HAFS, Yongin-si, Republic of Korea*

[2] *Department of Computer Science and Engineering, Korea University, KU, Seoul, Republic of Korea*

Abstract. MERS (Middle East Respiratory Syndrome) is a worldwide disease these days. The number of infected people is 1038(08/03/2015) in Saudi Arabia and 186(08/03/2015) in South Korea. MERS is all over the world including Europe and the fatality rate is 38.8%, East Asia and the Middle East. The MERS is also known as a cousin of SARS (Severe Acute Respiratory Syndrome) because both diseases show similar symptoms such as high fever and difficulty in breathing. This is why we compared MERS with SARS. We used data of the spike glycoprotein from NCBI. As a way of analyzing the protein, apriori algorithm, decision tree, SVM were used, and particularly SVM was iterated by normal, polynomial, and sigmoid. The result came out that the MERS and the SARS are alike but also different in some way.

1 Introduction

Middle East respiratory syndrome (MERS) was first identified in Saudi Arabia since an outbreak of new flu in 2012. The respiratory infection is caused by MERS coronavirus (MERS-CoV), case of which have been reported in 25 countries including Saudi Arabia, Egypt, United Kingdom, United States, and South Korea. 1,211 infections and 492 deaths were reported so far, which means about 40% of those who infected die of the disease. Since no specific vaccine and treatment of the virus were known as of 2015, the comparison of the virus with SARS coronavirus (SARS-CoV) has largely increased in the importance

1.1 SARS CoV

Severe Acute Respiratory Syndrome (SARS), a respiratory disease first appeared in China and spread out over the countries such as Singapore, Vietnam since 2002. Large number of both human and animals were affected severely by SARS for more than a decade. According to WHO, 8,096 infections and 774 deaths were reported in 2003[1]. Fever above 38°C is the initial symptom of SARS, which is followed by shortness of breath. SARS is spread out by SARS-coronavirus (SARS-CoV), which is a positive and single-stranded RNA virus. It has diverse structure of proteins, which include RNA replicase (1A, 1B), spike glycoprotein(S), small envelope glycoprotein(E), membrane glycoprotein(M), nucleocapsid phosphoprotein(N). Each protein has different roles, all of which are essential in various steps of virus life cycle [2].

1.2 MERS CoV

MERS (Middle East respiratory syndrome) causes lethal respiratory diseases. MERS is occurred by MERS-CoV, the single stranded RNA virus[3]. MERS is known to have come from camels, the host of the virus, but the exact route is not unfolded yet. Although MERS-CoV is coming up in endemic low level of threat, the possibility of MERS-CoV evolving to raise infectivity between human beings should not be overlooked. Corona virus is single stranded RNA virus. To enter host cells, MERS-CoV attaches to dipeptidyl peptidase 4, the receptor. And there is a process of fusion of the virus and the cell, using Protease cleavage of the S protein. Then the virus inserts the RNA to cytoplasm. The RNA of the virus then transcript and replicate on double membrane vesicles and other membranous structures, which are from the endoplasmic reticulum. Transcription of the seven subgenomic mRNA(messenger RNA)s happens through negative-strand subgenomic RNA intermediates. Subgenomic RNAs are 3′ co-terminally nested and they are joined to a common leader encoded at the 5′ end of the genome. The RNA of the virus is encapsidated in the N protein and is transported to the endoplasmic reticulum Golgi intermediate compartment (ERGIC), the site where they assembles. Viral RNA encapsidated in the N protein then buds into vesicles lined with the S, M, and E proteins. Vesicles are then transported to the cell surface before releasing [4].

[a] heresphill@gmail.com

1.3 Spike glycoprotein

Spike glycoprotein(S protein) is one of common structural proteins of coronavirus, belonging to class I fusion protein [5]. It is arranged on the surface of the virus forming a unique "corona", or crown-like appearance. The ectodomain of all coronavirus includes common two types of subunits, which are S1 domain and S2 domain. S1 domain, which is responsible for receptor binding, contains two subdomains, N-terminal subdomain and C-terminal subdomain. Both subdomains can function as receptor binding domains (RBDs) for host cells. 2 heptad repeats (HR1, HR2) can be found in S2 domain, which is responsible for virus-cell membrane fusion. The coronavirus infection begins with binding between RBD in S1 domain and receptor of the host cell, followed by viral fusion and entry resulted by the formation of fusogenic core between the HR1 and HR2 regions in S2 domain [6].

2 Method and experiment

2.1 Decision tree

Decision Tree is one of many data mining methods that are popular in many fields, including machine learning and statistical analysis. The algorithm can only classify category type structure inputs, which means in order to classify numerical value, it must first go through a discretization process to categorize the values [7]. By asking questions associated with data items, it uses a process of greedy algorithm, which branches down the original source into subsets, and shows the result of classification in a form of divided source. One of the pros of using decision tree algorithm is that it shows the result in a form of tree, which helps researches to analyze the result more easily [8].

2.2 Apriori algorithm

Apriori algorithm is famous for its simplicity though it is very effective in finding association rules between items. The algorithm takes two-step progress, which is finding large itemset and extracting all association rules which have confidence going over the minimum confidence [9]. In order to find large itemset, Apriori algorithm first calculates the support of each and every item sets that can be made, which is calculated in a formula of P(A).(Where A is an item set.) After calculation, it extracts every item set that has more support than the minimum support that we first set. It repeats the progress until it finds no item set that has higher support than the minimum support, and eventually creates a large itemset [10]. Then it calculates the confidence of each and every association rules that can be made in a large itemset, which is calculated in a formula of $P(A \mid B)$. (Where the association rule is $B \Rightarrow A$.) After calculation, it extracts every association rule that has more confidence than the minimum confidence that we first set. Thanks to its simplicity and effectiveness, this algorithm is widely used even to these days [11].

2.3 SVM (Suppport Vector Machine)

SVM algorithm is an initial of "Support vector machines". Also it is a supervised learning model with associated learning algorithms that analyze data and recognize patterns, used for classification and regression analysis. Now let us introduce the system of this algorithm [12]. First this algorithm makes a "Maximum margin hyper plane" to separate group into two parts, and let's say group A's vector is A vector(point) and group B's vector is B vector. And, let's connect these vector(points) and when a polygon is made, we call this convex hull. The important thing is the vector(point) which is outside the polygon. We say the line and the side that can separate the group "hyperplane". To test SVM's validity, we compare Accuracy rate. If Accuracy rate is high, two parts are similar. If not, two parts are very different[13].

2.4 Experiment

In our research, we used SARS and MERS Spike glycol protein data. We obtained SARS (DQ412574.1) and MERS(KP236092.1) Spike glycol protein sequence data in NCBI. Next, we performed experiment with decision tree algorithm first. In this experiment, we deterred MERS as class 1, SARS as class 2 and performed 10-Fold validation test in 7, 9, 13 window. In 10-Fold validation test, 9 data is used to make algorithm proper, and 1data is used to classify. Then, we conducted another experiment with apriori algorithm. We conducted experiment in 7, 9, 13 window. After experiment, we extracted rules from results. Finally, we conducted other experiment with SVM. We conducted this experiment in Normal, Polynomial, and sigmoid method. With 10-fold validation test, we used 8 data to make algorithm proper, and we used 2 data to classify

3 Results

3.1 Decision tree

Table 1. Results under 7 window

Class	Rule	frequency
Class1	Pos9=H	.75
Class2	Pos9=H	.923

Table 2. Results under 9 window

Class	Rule	frequency
Class1	Pos5=H	.75
Class2	Pos1=H	.75

Table 3. Results under 13 window

Class	Rule	frequency
Class1	Pos7=N	.8
Class2	Pos7=I	.818

In results under 7window, rules that frequency is higher than .75, rule-designating amino acids are quite similar. In this result, we could find that SARS and MERS are very similar viruses.

In 9window and 13window results, there are quite difference between two viruses. Comparing these differences, there is possibility that amino acids like isoleucine, asparagine, arginine can be key factor of making differences between two viruses

3.2 Apriori algorithm

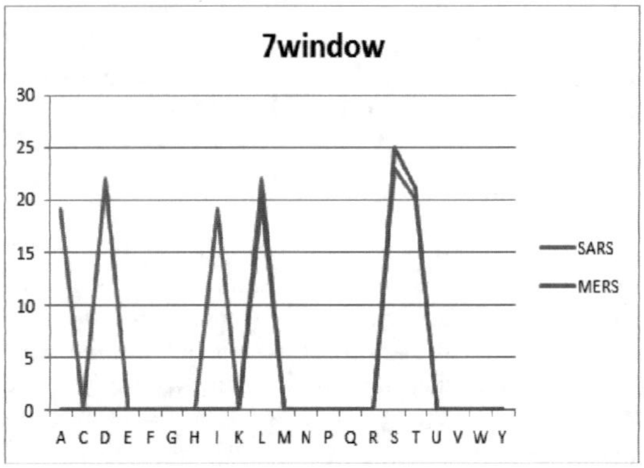

Figure 1. Results under 7window

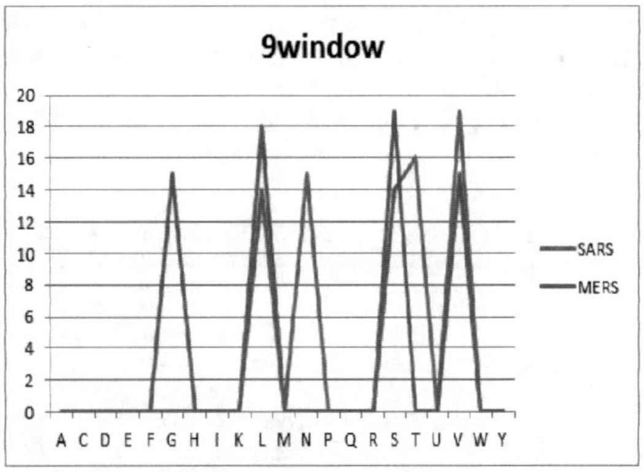

Figure 2. Results under 9window

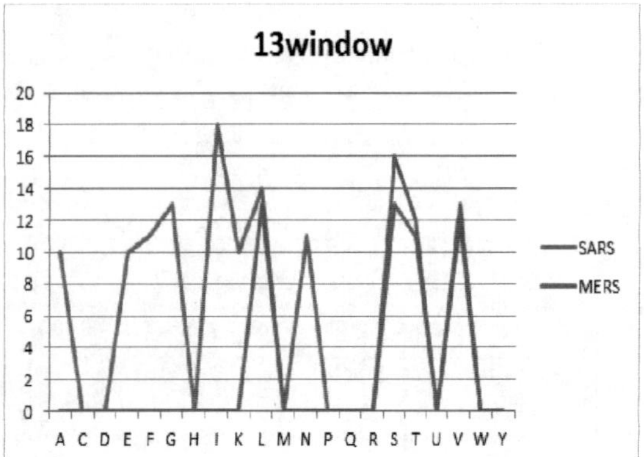

Figure 3. Results under 13window

As we can see in results by apriori algorithm, there is high similarity between SARS and MERS. However, there are some differences. First, as we can see in Figure 1 and Figure 3, amino acid sequence is differ from isoleucine between two viruses.

Next, in Figure 2 and Figure 3, we can find that asparagine is key factor that differ MERS from SARS. Comparing decision tree results under 13 window to apriori results under 13 window, we could find high similarity.

That is, isoleucine and asparagine are main factors that differentiate MERS from SARS

3.3 SVM

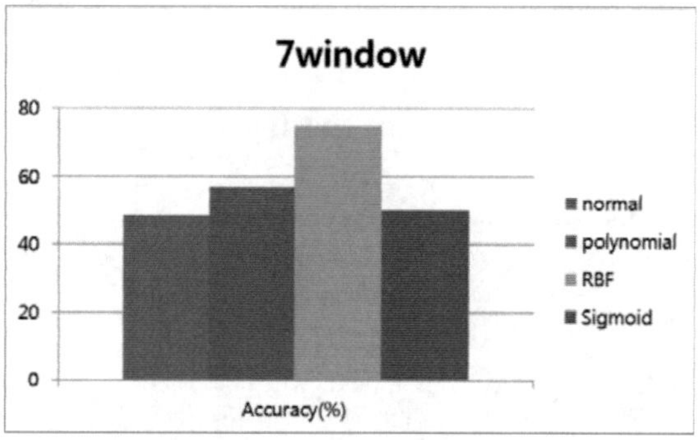

Figure 4. Results under 7window

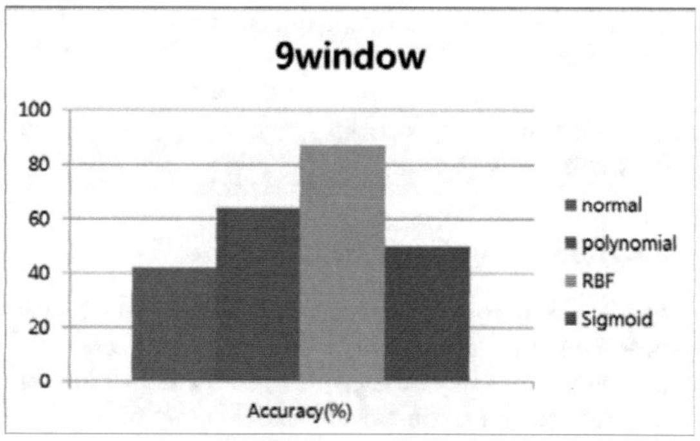

Figure 5. Results under 9window

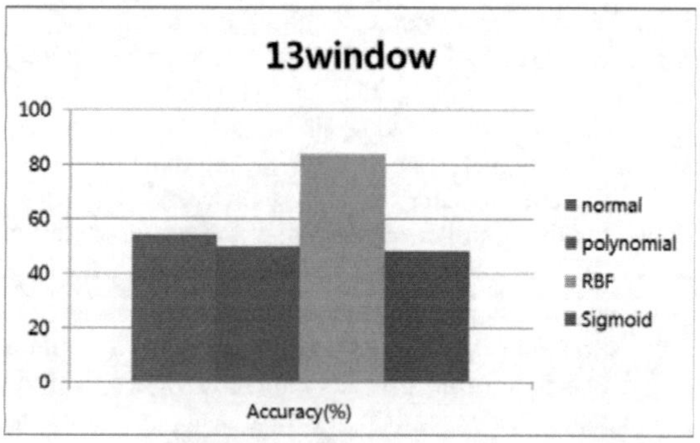

Figure 6. Results under 13window

In Normal SVM experiment, we can find that Accuracy rate is quite low. This means that there is obvious difference between MERS and SARS. Similarly, as we can see in Polynomial SVM experiment and sigmoid SVM experiment, we can find that other methods are showing low Accuracy rate. These results are showing that MERS and SARS is quite difficult to be distinguished by linear classifying process. However, different with those linear classifying process, RBF algorithm, which is non-linear classifying process, showed some positive results.

As we can look in 7 window, 9 window, 13 window results, the accuracy of RBF algorithm results are quite high. They are even higher than accuracy 75%. These results are showing that MERS and SARS viruses are very similar viruses. However, by non-linear classifying process, MERS and SARS viruses can be distinguished easily. In conclusion, MERS and SARS viruses are quite different viruses at all.

4 Conclusion

This paper applied the computational program, apriori algorithm, decision tree, and SVM. To conduct our experiment, we used Spike glycoprotein of MERS and SARS. To analyze this, we used three different algorithms. Apriori algorithm is algorithm that extracts frequent rules from given dataset. With decision tree, we can classify two viruses. With SVM, we found that MERS and SARS are truly different viruses.

MERS virus is very critical and epidemic disease. However just like our experiment, MERS is different from SARS. In order to block the virus from spreading out and killing people, we performed experiment to analyze key factor of MERS virus.

As we can find, rule extracted from apriori algorithm showed that amino acid Isoleucine and Asparagine are the main factors of MERS. For further research we are going to conduct experiment with other proteins except Spike glycoprotein.

References

1. Peiris, J. S. M., *et al.* Clinical progression and viral load in a community outbreak of coronavirus-associated SARS pneumonia: a prospective study, The Lancet, **361**, 9371, 1767-1772 (2003)
2. Seto, W. H., *et al.* Effectiveness of precautions against droplets and contact in prevention of nosocomial transmission of severe acute respiratory syndrome (SARS), The Lancet, **361**, 9368, 1519-1520 (2003)
3. de Groot, Raoul J., *et al.* Middle East respiratory syndrome coronavirus (MERS-CoV): announcement of the Coronavirus Study Group, Journal of virology, **87**, 14, 7790-7792 (2013)
4. Assiri, Abdullah, *et al.* Hospital outbreak of Middle East respiratory syndrome coronavirus. New England Journal of Medicine, **369**, 5, 407-416 (2013)
5. Simmons, Graham, *et al.* Characterization of severe acute respiratory syndrome-associated coronavirus (SARS-CoV) spike glycoprotein-mediated viral entry. Proceedings of the National Academy of Sciences, **101**, 12, 4240-4245 (2004)
6. Qian, Zhaohui, Samuel R. Dominguez, and Kathryn V. Holmes. Role of the spike glycoprotein of human Middle East respiratory syndrome coronavirus (MERS-CoV) in virus entry and syncytia formation. PloS one, **8**, 10, e76469 (2013)
7. Borgelt, Christian, and Rudolf Kruse. Induction of association rules: Apriori implementation, Compstat. Physica-Verlag HD (2002)
8. Borgelt, Christian, Recursion Pruning for the Apriori Algorithm, FIMI (2004)
9. Friedl, Mark A., and Carla E. Brodley, Decision tree classification of land cover from remotely sensed data, Remote sensing of environment, **61**, 3, 399-409 (1997)
10. Magerman, David M. Statistical decision-tree models for parsing. Proceedings of the 33rd annual meeting on Association for Computational Linguistics, Association for Computational Linguistics (1995)
11. Safavian, S. Rasoul, and David Landgrebe. A survey of decision tree classifier methodology, IEEE transactions on systems, man, and cybernetics **21**, 3, 660-674 (1991).
12. Schüldt, Christian, Ivan Laptev, and Barbara Caputo. Recognizing human actions: a local SVM approach. Pattern Recognition, 2004. ICPR 2004. Proceedings of the 17th International Conference on. Vol. 3. IEEE (2004)
13. Furey, Terrence S., *et al.* Support vector machine classification and validation of cancer tissue samples using microarray expression data, Bioinformatics, **16**, 10, 906-914 (2000)

MATEC Web of Conferences **49**, 09001 (2016)
DOI: 10.1051/matecconf/20164909001

Reliability of Using Elastic Modulus for Non- Homogeneous Materials

Rimsha Khan [1], Zartasha Mustansar [1, a]

[1]*Research Center of Modeling and Simulations (RCMS), National University of Sciences and Technology (NUST), Sector H-12, Islamabad, 44000, Pakistan*

Abstract. This paper investigates the literature available for derivation of the elastic modulus (E) as employed specifically in non-homogenous materials. There is a big pool of E-values in the literature which user can pick and use. However their reliability is still a question. This study will therefore focus on the values of elastic modulus acquired from the secondary databases and their reliability in numerical methods like Finite Elements (FE). This study uses E-values obtained from bones, muscles and tissues. Bone is a composite material with a very intricate internal geometry. Acquiring the elastic modulus values for irregular geometries like bones is not a straight forward process. This study therefore compares the different studies provided in literature for derivation of elastic properties of bones from different vertebrates and investigates the reason of variations. The study infers that populating all the values of elastic modulus without a standard procedure, creates a level of ambiguity that bars the understanding of the use of modulus of elasticity where insightful procedures are in question for example finite element analysis of sensitive materials . Therefore there is a need to take great care in deriving the values of E.

1 Introduction

The Elastic Modulus (E) is a mechanical property of solids in which the force is exerted on a body to cause the deformation [1]. Elastic modulus (E) defines the relationship between stress in which force acts per unit area whereas strain signify deformation in a material [2]. While dealing with mechanics of biological materials the structural properties and mechanical properties are in major focus. Structural properties show a relationship between the force and deformation while the mechanical properties express the behavior of material in terms of stress-strain relationship in the materials [3].

The strength of a material will always be different compared to the corresponding modulus of elasticity or stiffness except bending mode of loading. It is known that stiffness is the property of a material that represents the ability of a material to resist the deformation while it has been observed that a stiffer material has a higher elasticity to deform [4]. The need to understand why elastic modulus value could be misleading is due to the large variation of its values. There is a big factor of ambiguity in literature as to what method of testing holds true for a specific value of elastic modulus.

In a linearly elastic material, the Young's Modulus describes how a material responds to stress. However, if we consider a microscopically refined level, it is similar to a spring constant. Conceptually, atoms in a solid are bonded to other atoms. If you force two atoms apart, there will be a force that attempts to restore them to their equilibrium distance. This is similar to having two wooden blocks connected to a spring and then pulling them apart. As a result, the blocks will jump back together if you let go. The restoring force depends on what the spring is made of, so the Young's modulus describes how "stiff" the bonds of a solid are relative to the material itself.

Thus, in archiving values for the modulus of elasticity for non-homogenous materials, materials should be sensibly characterized. Materials are of two types' ideal materials and non-ideal materials. Ideal materials are isotropic and homogenous which shows that their properties are same in all directions while anisotropic and non-homogenous material varies in all directions [5]. This study will include all the non-homogenous material data for bone, muscles and tissue to investigate the cause of variation in the values of elastic modulus.

[a] Corresponding author: zmustansar@rcms.nust.edu.pk

2 Background

A Bone is one of the major composite materials which consist of both fluid and solid phases [6]. Bone is made up of two main solid phases; organic and inorganic phase which provides hardness to the bone. The organic phase provides flexibility to the bone while the inorganic material of the bone gives its resilience [7]. In any long bone ,the cortical bone is roughly four times the mass of cancellous bone [8]. The porosity of cortical bone ranges from 5 to 30% while cancellous bone's porosity ranges from 30-90%. The bone porosity is not fixed and it can be changed in response to the altered load, or any disease and also other factors like aging [9]. The mechanical properties of biomaterial are difficult to predict.

This study will therefore focus on the values of elastic modulus acquired from the secondary databases and their reliability. Various studies are conducted to acquire value of elastic modulus using different types of testing methods. Studies including [1], [3] where deformations based acquisition of bone's properties employing theoretical models are predicted; the accuracy is a function of material properties itself. To-date these values in the literature is a matter of question.

Recent studies have shown new ways of predicting the accuracy of the biomaterials as in case of bones. Some of which considered the entire surface of the bone for deformation [10] and others entirely different procedures. In the next section, mechanics of bone including various types of test to derive elastic modulus is explained briefly.

3 Mechanics of bone

This section will cover the mechanics of biomaterials (especially bones) in correspondence to the elastic modulus values. A bone is a term used for hard and rigid structure of the vertebral skeleton system, that gives shape and support to the body and it also protect the body [11]. Composition, structure and mechanical properties of bone may continuously vary or in discrete steps from one part to another part [10]. The arrangement of constituents of bones and its classified structure give rise to a very complex material type structure that cannot define the single significance for particular mechanical property. Next sections explain some basic types of testing methods to find values of elastic modulus.

Stress and Strain are the properties of the materials such that, that when an external force is applied the structure deforms. Stress is defined as internal force per unit area [12], on the contrary strain is the change in shape or size of an object due to the outwardly applied forces to the object [13]. Stress is much more complex entity in more rigorous situations and the units of stress is Pascal (Pa) which is equal to the one Newton per meter square ($1 N/m^2$)[14].

Among the various forms of stresses, tensile stress is the type of stress in which the length of material extends and it act as normal to the stressed area (Fig. 1). Compressive stress is the type of stress that have tendency to compress or shorten the length of the material, it also acts normal to the stressed area and shear Stress is the type of stress that has tendency to act in the planes of stressed area at the right angles [15, 16].

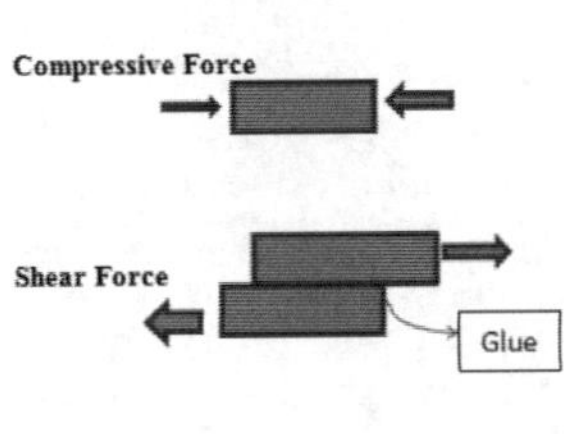

Figure 1. Illustration of tension, compression and shear forces.

In all these types of testing methods, it is easy to develop a stress-strain curve, which can be eventually used to derive the values of elastic modulus.

4 Derivation of elastic modulus using different loading tests

Most studies of the mechanical testing of bones depend upon `the in vivo and in vitro experiments. Torsional and impact loading are less frequently used. Our investigation in this study will only focus on the most commonly used mechanical tests in literature. Among the traditional testing methods like bending, compression and tension tests, finite element method recently, is extensively being used to determine the modulus of elasticity in biomaterials [17].

Bending test is mainly used to illustrate the mechanical behavior of the biomaterials [18-21]. This test however measures the coarse value of Young's modulus or Elastic modulus (E) from the geometry of the loading device in some cases [22] and at the same time restrains toughness of the material [23]. One of the studies show the value of elastic modulus of the compact bone after applying the mechanical test of bending is 115 GPa [24] for apatite which is over-estimated in our knowledge while on the other hand, for the muscles that are attached to this consumes a value of 40.3GPa from bending [9].

Torsion is another form of basic loading mode, where forces tend to twist the object along the longitudinal axis [25]. The intricate torsion forces can be used to acquire the values of elastic modulus by generating stress-strain curve. In tension, the two forces act on an object and try to pull the object in opposite direction. As a result the object is stretched and become thinner and longer than the original one [19].

Bones respond intermediately to the tension forces. This is the most basic method to obtain value Young's modulus (again) by creating a load-force curves. It is possible to do tension test on muscles as well in order to obtain stiffness. Muscles respond to the tension forces called tension load. They either contract that pulls the tendons at both ends. This cause little stretch in the muscles and length of the tissues also change [26].

Various studies show several values of elastic modulus in muscles, bones and tissues [27]. Zajac [28] proposed the value of elastic modulus as 80 GPa for a muscle and for the compact human bone the value of elastic modulus was reported as 0.053 GPa by [29]. Continuing Doblare's work [29] regarding modeling of bone tissue fracture and healing, some contradictory values of elastic modulus (E) were also reported for compact human bone [9].

This difference primarily is due to the different loading tests in both data sets. Compression however, involves forces that push the object along the axis and makes it thicker and bulge out as a result the material length is shortened [30]. Tension and compression forces both act as axial stresses because they both act on structure longitudinal axis. The value of elastic modulus acquired by Reilly and Burstein using compression is reported as 0.131GPa on a human compact bone [31].

Other compression tests including tests on cortical femoral bone reports E-value of 2.065GPa [27] while the human trabecular bone with an E-value of 1.904GPa [32] in the same axial direction using different bone-types. Another study investigates the compressive strength of the cancellous bone as 20GPa [5]. Similarly, numerical methods such as Finite Element Method (FEM), which is one of the most widely used methods to assess the strength and quality of bone in terms of stiffness; has been in practice to identify the behavior of bone, for the past 40 years [3]. Various studies have been conducted for the derivation of E-values and to manipulate the mechanical behavior of bones, muscles and tissues [33-35]. It is not possible to cite all the relevant references using FEM, in this short investigation paper however one of the studies reports E-value as 18.4GPa [36] which is considered to be the most widely used E-value for cancellous bone.

5 Results and conclusions

Many studies show that behavior of bone is affected by different mechanical properties [5]. After careful investigation, we have identified different studies which are using various kinds of bones employing different scheme of methods; in order to derive values for elastic modulus. In total, the data was roughly pulled out from studies reported in literature. It should also be noted that the investigation carried out in this paper is based on twenty study subjects (Fig. 2) only, which is not a very big number and that's probably one of the limitations of this study.

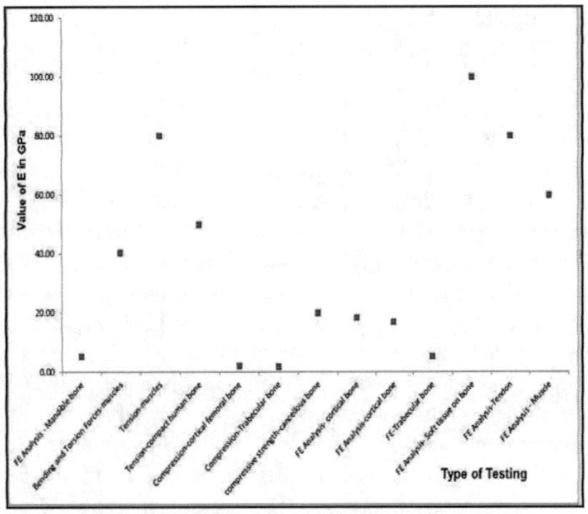

Figure 2. Scatter of Young's modulus values for different types of loading tests.

Each study focused on the derivation of elastic modulus. It is observed, that there is no consistency in the values of elastic modulus, throughout the literature. Perhaps one reason might be the use of different loading and testing methods to acquire values of elastic modulus. Additionally, different authors use bones from different species and different parts. For example an elastic modulus for human compact bone from the femur will yield different value compared to the elastic modulus of the human compact bone from the humerus.

There is a significant need to pay attention to this area where values of elastic modulus is of great importance (especially) for the applications which requires sensitive procedures. In the field of civil engineering where elastic modulus of concrete holds much importance because this mechanical parameter reflects the ability of the concrete to deform elastically. For example, in pre-stressed concrete structures, elastic shortening of pre-stressed concrete is one of the main factors contributing to pre-stress loss.

Experiments shows that Young's modulus value also increases if the bone starts to mineralize more due to high mineral content and collagen fibers [7]. The collagen fibers tend to increase the value of elastic/Young's modulus due to high value of stiffness [37].Usually in the literature the bone's elasticity is ranged between 15-25 GPa [38] which is one third of a metallic aluminum [39]. The bone's elasticity is important because, it gives the bone ability to dissipate energy during deformation phenomena's. Methods like Finite Elements, which requires the value of elastic modulus as one of the major inputs predicts results of simulation on the basis of E-value that is used as an input. If one value goes wrong the whole simulation will result into an inaccurate and ambiguous situation. Apparently, the criticality of this basic factor is under estimated and it might result into the situations which do not relate to any real physical phenomena's at all.

Therefore, there is a need to look into the importance of methods used to derive elastic modulus of various materials and the steps followed, so that a consistency can be maintained in the literature housing the values of Young's modulus over a wide range.

References

1. D. Burr, C. Milgrom, D. Fyhrie, M. Forwood, M. Nyska, A. Finestone, et al., In vivo measurement of human tibial strains during vigorous activity, Bone, **18**, 405-410 (1996)
2. C. Milgrom, A. Finestone, A. Hamel, V. Mandes, D. Burr, N. Sharkey, A comparison of bone strain measurements at anatomically relevant sites using surface gauges versus strain gauged bone staples, Journal of biomechanics, **37**, 947-952 (2004)
3. M. J. Silva, M. D. Brodt, W. J. Hucker, Finite element analysis of the mouse tibia: Estimating endocortical strain during three - point bending in SAMP6 osteoporotic mice, The Anatomical Record Part A: Discoveries in Molecular, Cellular, and Evolutionary Biology, **283**, 380-390 (2005)
4. H. Gao, Application of fracture mechanics concepts to hierarchical biomechanics of bone and bone-like materials, International Journal of Fracture, **138**, 101-137 (2006)

5. L. J. Gibson, The mechanical behaviour of cancellous bone, Journal of biomechanics, **18**, 317-328 (1985)
6. S. C. Cowin, Bone mechanics handbook, (2001)
7. P. D. Delmas, R. P. Tracy, B. L. Riggs, K. G. Mann, Identification of the noncollagenous proteins of bovine bone by two-dimensional gel electrophoresis, Calcified tissue international, **36**, 308-316 (1984)
8. A. Sharir, M. M. Barak, R. Shahar, Whole bone mechanics and mechanical testing, The Veterinary Journal, **177**, 8-17 (2008)
9. J. D. Currey, The effect of porosity and mineral content on the Young's modulus of elasticity of compact bone, Journal of biomechanics, **21**, 131-139 (1988)
10. S. Suresh, Graded materials for resistance to contact deformation and damage, Science, **292**, 2447-2451 (2001)
11. D. G. Steele, C. A. Bramblett, The anatomy and biology of the human skeleton: Texas A&M University Press, (1988)
12. Y.-c. Fung, P. Tong, Classical and computational solid mechanics vol. 1: World scientific, (2001)
13. H. Davies, The timing and distribution of strains around the surface of the midshaft of the third metacarpal bone during treadmill exercise in one Thoroughbred racehorse, Australian veterinary journal, **83**, 157-162 (2005)
14. C. Bitsakos, J. Kerner, I. Fisher, A. A. Amis, The effect of muscle loading on the simulation of bone remodelling in the proximal femur, Journal of biomechanics, **38**, 133-139 (2005)
15. A. Palevski, I. Glaich, S. Portnoy, E. Linder-Ganz, A. Gefen, Stress relaxation of porcine gluteus muscle subjected to sudden transverse deformation as related to pressure sore modeling, Journal of biomechanical engineering, **128**, 782-787 (2006)
16. Y. Zheng, A. F. Mak, Effective elastic properties for lower limb soft tissues from manual indentation experiment, Rehabilitation Engineering, IEEE Transactions on, **7**, 257-267 (1999)
17. A. G. Robling, D. B. Burr, C. H. Turner, Recovery periods restore mechanosensitivity to dynamically loaded bone, Journal of Experimental Biology, **204**, 3389-3399 (2001)
18. C. T. Rubin, L. E. Lanyon, Regulation of bone mass by mechanical strain magnitude, Calcified tissue international, **37**, 411-417 (1985)
19. P. Ammann, R. Rizzoli, Bone strength and its determinants, Osteoporosis International, **14**, 13-18, (2003)
20. S. J. Warden, J. A. Hurst, M. S. Sanders, C. H. Turner, D. B. Burr, J. Li, Bone adaptation to a mechanical loading program significantly increases skeletal fatigue resistance, Journal of bone and mineral research, **20**, 809-816 (2005)
21. N. E. Lane, W. Yao, M. Balooch, R. K. Nalla, G. Balooch, S. Habelitz, et al., Glucocorticoid - Treated mice have localized changes in trabecular bone material properties and osteocyte lacunar size that are not observed in placebo - Treated or estrogen - deficient mice, Journal of bone and mineral research, **21**, 466-476 (2006)
22. S. Judex, X. Lei, D. Han, C. Rubin, Low-magnitude mechanical signals that stimulate bone formation in the ovariectomized rat are dependent on the applied frequency but not on the strain magnitude, Journal of biomechanics, **40**, 1333-1339 (2007)
23. E. Linder-Ganz, N. Shabshin, Y. Itzchak, and A. Gefen, Assessment of mechanical conditions in sub-dermal tissues during sitting: A combined experimental-MRI and finite element approach," Journal of biomechanics, **40**, 1443-1454 (2007)
24. J. Katz, The biophysical and biomechanical properties of bone, bone-mineral and some synthetic bone biomaterials, bulletin de la Société chimique de France, 514-518 (1985)
25. J. Nowinski, C. Davis, The flexure and torsion of bones viewed as anisotropic poroelastic bodies, International Journal of Engineering Science, **10**, 1063-1079 (1972)
26. R. B. Bach, D. Burr, N. A. Sharkey, Skeletal tissue mechanics: Springer Science & Business Media, (2013)
27. J. C. Lotz, T. N. Gerhart, W. C. Hayes, Mechanical properties of metaphyseal bone in the proximal femur, Journal of biomechanics, **24**, 317-329 (1991)
28. F. E. Zajac, Muscle and tendon: properties, models, scaling, and application to biomechanics and motor control, Critical reviews in biomedical engineering, **17**, 359-411 (1988)
29. M. Doblaré, J. Garcıa, M. Gómez, Modelling bone tissue fracture and healing: a review, Engineering Fracture Mechanics, **71**, 1809-1840 (2004)
30. D. M. Cullinane, T. A. Einhorn, Biomechanics of bone, Principles of bone biology, **1**, 17-32 (2002)
31. D. T. Reilly, A. H. Burstein, The elastic and ultimate properties of compact bone tissue, Journal of biomechanics, **8**, 393-405 (1975)
32. S. Pietruszczak, D. Inglis, G. Pande, A fabric-dependent fracture criterion for bone, Journal of biomechanics, **32**, 1071-1079 (1999)
33. M. Zhang, Y. Zheng, A. F. Mak, Estimating the effective Young's modulus of soft tissues from indentation tests—nonlinear finite element analysis of effects of friction and large deformation, Medical engineering & physics, **19**, 512-517 (1997)
34. T. J. Hughes, The finite element method: linear static and dynamic finite element analysis: Courier Corporation, (2012)
35. J.-Y. Rho, L. Kuhn-Spearing, P. Zioupos, Mechanical properties and the hierarchical structure of bone, Medical engineering & physics, **20**, 92-102 (1998)
36. T. Gardnera, T. Stoll, L. Marks, S. Mishra, M. K. Tate, The influence of mechanical stimulus on the pattern of tissue differentiation in a long bone fracture—an FEM study, Journal of biomechanics, **33**, 415-425 (2000)
37. V. K. Goel, S. A. Ramirez, W. Kong, L. G. Gilbertson, Cancellous bone Young's modulus variation within the vertebral body of a ligamentous lumbar spine—application of bone adaptive

remodeling concepts, Journal of biomechanical engineering, **117**, 266-271 (1995)
38. B. Van Rietbergen, R. Huiskes, F. Eckstein, and P. Rüegsegger, Trabecular bone tissue strains in the healthy and osteoporotic human femur, Journal of Bone and Mineral Research, **18**, 1781-1788 (2003)
39. B. Van Rietbergen, A. Odgaard, J. Kabel, R. Huiskes, Direct mechanics assessment of elastic symmetries and properties of trabecular bone architecture, Journal of biomechanics, **29**, 1653-1657 (1996)

AUTHOR INDEX